A VIEWER'S
HALLEY'S COMET

The appearance of Halley's comet has astonished, fascinated—even terrified—man for centuries. One of the most colorful observers was Edmund Halley (1656–1742), the man who first charted and named the comet. Did you know that:

- Edmund Halley was an inventor, navigator, sea captain, explorer, deep-sea diver—and a college dropout who received his M.A. from Oxford only after the King of England pulled some strings.

- Most people pronounce his name "Hay-lee," probably because of the pop group Bill Hayley and the Comets. Actually, Edmund Halley himself may have pronounced his name "Haw-ley" or "Hally" (rhymes with "Sally").

- One of Halley's closest friends was a brilliant but disorganized professor. After much hounding, Halley finally talked his friend into writing a series of books. As a result of Edmund Halley's persistence, the world has Sir Isaac Newton's masterpiece, the *Principia*—containing the basic laws of physics!

A VIEWER'S GUIDE TO HALLEY'S COMET

MATTHEW HART

PUBLISHED BY POCKET BOOKS NEW YORK

Another *Original* publication of POCKET BOOKS

POCKET BOOKS, a division of Simon & Schuster, Inc.
1230 Avenue of the Americas, New York, N.Y. 10020

ISBN: 0-671-49841-X

First Pocket Books printing September, 1985

10 9 8 7 6 5 4 3 2 1

POCKET and colophon are registered trademarks
of Simon & Schuster, Inc.

Printed in the U.S.A.

For Gregory, Peter, Robert, Michèle,
and Mary-Lynne,
my brothers and sisters

Acknowledgments

The author thanks the following for their kind permission to quote passages from other works:

The excerpt from *Worlds in Collision* by Immanuel Velikovsky, copyright 1950 by Immanuel Velikovsky, and the excerpt from *Edmund Halley: Genius in Eclipse* by Colin Ronan, copyright 1969 by Colin Ronan, are reprinted by permission of Doubleday and Company, Inc. The excerpts from the articles by Paul Hoffman and Wallace Tucker which appeared in *Science Digest* are reprinted by permission of the authors.

Two quotations from the *Correspondence of Isaac Newton* appear by kind permission of the Royal Society. The long passage from George Chetwynd Griffith's 1894 classic *Olga Romanoff* is quoted by permission of Hyperion Press, Inc.

Taylor and Francis Limited has permitted the use of several quotations taken from E. F. MacPike's *Correspondence and Papers of Edmund Halley*.

Many of the translations of early historians and observers are taken from the *Flammarion Book of Astronomy* and are here reprinted by permission of George Allen and Unwin.

Permission to quote from their monograph *Project Icarus*, edited by Louis Kleiman, comes from M.I.T. Press, Cambridge, Massachusetts, and is gratefully acknowledged.

Contents

Halley's Comet Countdown

1059 B.C.	Earliest recorded observation.
12 B.C.	Comet appears from August 26 to October 20. Contrary to popular belief, Halley's was *not* the star of Bethlehem.
A.D. 66	Comet described as a "sword hanging in the sky." It was this appearance that was credited by contemporaries with responsibility for the destruction of Jerusalem.
A.D. 684	Earliest recorded drawing of the comet. Later published in the *Nuremburg Chronicles* of 1493.
A.D. 1066	Comet appears to mark the invasion of England by William the Conqueror.
A.D. 1301	Comet seen by the great painter Giotto di Bondone, who awarded it the role of the star of Bethlehem in his painting the *Adoration of the Magi*.

HALLEY'S COMET COUNTDOWN

A.D. 1456	Pope Calixtus calls the comet an "agent of the devil."
A.D. 1682	This is the apparition observed by Edmund Halley.
A.D. 1759	Comet returns according to Halley's prediction, proving that comets orbit the Sun. The great comet is named for Halley.
A.D. 1835	Comet appears at birth of Mark Twain.
A.D. 1910	Comet appears at death of Mark Twain.
A.D. 1985	
November	First close approach; comet may be viewed through small telescopes and binoculars.
December	May be viewed by naked eye in evening.
A.D. 1986	
January 1–20	Naked-eye viewing possible where skies are not ruined by light pollution. Early evening.
February 9	Perihelion (closest point to the Sun).
February 20–March 15	Reappears before dawn. Naked-eye viewing possible. Tail starts to lengthen appreciably.
March 15–25	Best viewing for those living above latitude 35 north—roughly, above a line drawn from Cape Hatteras, off the North Carolina Coast, to San Luis Obispo, above Los Angeles. Look in the southeast for a few hours before dawn. Tail closest to greatest length.
April 10–11	Comet closest to Earth, but also farthest south. Best viewing below the equator.

April 12–26	Comet moves rapidly north. Tail shortens and comet becomes dimmer. Visible throughout night. Moon becomes a problem.
April 26– May 4	Last naked-eye viewing. Visible throughout most of night.
May–August	Visible through small telescopes until lost in the glare of the Sun.
A.D. 1987	Halley's comet becomes a memory . . . until 2061!

A VIEWER'S GUIDE TO
HALLEY'S COMET

a fraction of the mass it looks to be, then the passage of

INTRODUCTION

On Christmas night, 1758, a Saxon farmer named Johann Georg Palitzsch sat in his fields scanning the winter sky. Palitzsch was an amateur astronomer, and more than a casual one, for he was searching the heavens with a seven-meter reflecting telescope that he had built himself. Many of the leading astronomers of the day were also searching through the sky that night, but Palitzsch saw first what they had all been seeking: Halley's comet. The event was momentous, proving at last that man could accurately predict the return of an exotic celestial visitor.

Ever since then, Halley's comet has been one of the most famous objects in the Solar System. Many of us have an elderly relative who can remember when Halley's made its last pass around the Sun in 1910 (or if we haven't got one of our own, we know where we can lay our hands on one). Get him or her to tell you about comet pills, or about the special helmets designed to protect the human head from plunging debris. Mind you, sometimes a little judicious fear is not such a wrongheaded thing. What if comets cause the plague?

In the summer of 1983 bubonic plague suddenly appeared in the American Southwest. In five states, thirty-five patients exhibited the fever, chills, headache, and swelling in the armpits, groin, or neck that mark sufferers. Six of them died, one of them a thirteen-year-old boy whose condition had been diagnosed as influenza.

That same year bubonic plague also struck the battlefields of Namibia, where South African troops fight the guerrillas who seek to wrest the vast, rich desert region from Pretoria's control. More than 500 cases of plague were reported in Namibia.

There is no reason to panic, of course, but there is at least the opportunity to reflect on the simultaneous appearance of this ancient menace in such widely separated parts of the world. And perhaps there is the opportunity also to wonder whether the theory propounded by Sir Fred Hoyle (discussed in Chapter 7) might be something more than well-fashioned tripe. In their theory, Sir Fred and a colleague maintain that diseases which attack man and animal alike fall to Earth from passing comets. They are serious about it. But judge for yourself. Just keep an eye on Halley's comet, due in 1985–86.

In 1948 Halley's reached aphelion—its farthest point from the Sun—and paused far out in the blackness 3,255,000,000 miles away. Ever since then it has been falling back toward the Sun, gaining a little more speed with each second that our star draws it nearer. As it drew close enough actually to be seen, the excitement among astronomers increased, and each team manning the world's great optical instruments was determined to be the first to spot the comet. The United States won. In the early-morning hours of October 16, 1982, graduate student David C. Jewitt and staff astronomer

G. Edward Danielson took seven pictures of the comet. They were using a sensitive electronic camera attached to the mighty 200-inch Hale telescope atop California's Mount Palomar. Since Halley's was the ninth comet to be either discovered or rediscovered (*recovered*, in astronomical terms) in 1982, it was designated 1982i by the International Astronomical Union.

There is an unavoidable peril in writing any book on an astronomical subject: new discoveries pour in so fast that you barely get the covers on the thing before someone is leafing through it and snickering at its quaintness. For that reason, I have tried to keep all discussion of current material on as sound a footing as possible, sticking to the experts. Here is an example.

IRAS is the acronym for the Infrared Astronomical Satellite, in effect an orbiting telescope, jointly operated by the United States, Britain, and the Netherlands. In 1983 two American astronomers trained IRAS on Vega, a star in the constellation Lyra and one of the brightest stars in the summer sky. As they zeroed in on Vega the astronomers realized that something strange was happening. They were picking up heat signals not just from the star itself . . . but from the region *around* the star. Vega, a much younger star than the Sun, seemed to be surrounded by orbiting matter, possibly an evolving solar system of her own.

The temptation to pile onto the bandwagon of speculation about Vega's orbiting material is hard to resist. But the conclusions about the nature of the material are thus far so tentative, so utterly speculative, that there seems little reason to bother right now. Vega will be thoroughly explored soon enough, and I for one can wait. This does not mean that the book avoids all speculation. Where would be the fun? The chapter on extraterrestrial life, for instance, is really elaborate self-indulgence. But in general I have tried to construct a book that comfortably situates the reader in the universe he shares with good old 1982i—Halley's comet.

On our way through the material, we encounter, from time to time, a crackpot. But what of it? Crackpots operate within the boundaries of an ancient tradition. About A.D. 150, the Greek philosopher Lucian of Samosata was writing about the adventures of voyagers whose ship was carried to the Moon on a waterspout. The crew found the Moon-King and the Sun-King at war over who would rule Jupiter. They also found archers astride giant fleas, and people with artificial genitals. Wisely, the Earthlings returned home.

Anyone interested in astronomy today will be grateful for the stirring and richly imaginative books by such men as Carl Sagan, Robert Jastrow, Isaac Asimov, and scores of other brilliant and thoughtful astronomers. I have managed to mention only a few of them in this book, but I am indebted to them all. And the standard proviso must apply: the visions are theirs; the mistakes are mine.

There are many reasons why we study comets, and why we will be spending so many millions of dollars to get a close look at Halley's. Comets are among the most primitive objects in the Solar System, and in studying them we are studying our own origins. Also, the physical mechanisms of comets tell us a lot about interplanetary and interstellar space. Most of all, there is our own boundless curiosity about the universe and the objects that travel even our own little corner of those trillion trillion trillion miles of rushing galaxies.

Besides, it's fun.

M.H.
March 1985

CHAPTER 1

In Terror of the Sky

One day, there can be little doubt, some wicked chunk of matter is going to come whistling out from between the planets and, unless we blow it away with our fancy zappers, bury under a few miles of fallout those of us it does not pulverize. If it hits an ocean, many of us will simply be sluiced into the sea. If it comes thumping ashore onto dry land, the lucky ones will be those standing directly beneath it. For them it will be quick. The rest will have the teeth shaken out of their heads and spend the last years of mankind's tenure on Earth groping about in the gloom of the great dust cloud such an impact will raise. Now, don't load the revolver just yet; the odds are that the collision won't threaten for a few thousand years!

The ancients, although for the wrong reasons, were probably wiser than their insouciant descendants when they watched the great comets plow across the skies above their towns or villages and shuddered in their hearts. They were passionate observers, those antecedents of ours. They knew the orderly motions of the stars and planets, knew they moved at the behest of the gods, knew the motions were predictable, if inscrutable. And then comes this blundering savage, storming into the regulated universe and smearing his tail of fire across the night.

The word *comet* comes from the Greek word *kometes,* meaning *hairy;* the Greeks called the comet *aster kometes,* or *hairy star,* although the Greeks themselves attributed the origin of the term *hairy* to the Egyptians.

It was probably the very transitoriness of comets that led our ancestors so insistently to tie their appearance in the sky to events below. Since the comets were so strange, so extraordinary and fleeting, anything important that happened while they were in the sky must seem connected to, and by extension caused by, the evil tongues of flame.

Some observers, more dispassionate than their fellows, believed that comets originated in the Earth as collections of gas. These gaseous pockets were then exhaled into the sky, where, upon reaching the level of the fiery sphere, they burst into flame. Still others believed that comets were the souls of great men leaving the planet that had been their home in life and moving to take their rightful place in the pantheon of gods and heroes. The Romans were convinced that the comet of 44 B.C. was the soul of Julius Caesar, who died that year. A god himself, they thought, Caesar would naturally make a fairly serious splash as he left to join all the other gods. Once installed in heaven, he would presumably take as keen an interest in the affairs of Rome as he had before moving from the manager's office into the boardroom.

Nevertheless, it was as an omen of disaster that most observers saw the appearance of a comet in the otherwise orderly heavens. That is why comets were so often depicted by ancient writers as taking dreadful shapes. They were described as representing every weapon of war in use: clubs, javelins, maces, battleaxes, swords, daggers; their colors were always lurid; they sometimes resembled decapitated heads dripping gore and held aloft by the hair in the grip of a vengeful fist. These are not the sorts of descriptions calculated to promote much in the way of cool reflection. And they didn't.

Not Just Another Job

The emperor Nero (reigned A.D. 54–68) was so certain that a comet appearing during his reign meant his death that he began to murder Roman aristocrats in order to appease the bloodthirsty comet and save his own soft skin. Maybe we shouldn't knock it; it worked. Nero lived, at least for another few years, and no doubt rewarded his court astrologer, one Babilus, for warning him about the comet.

Nero was not alone in being in a position to secure this kind of valuable advice. Right up until the seventeenth century it was common for monarchs to keep astrologers at court. But it wasn't always an enviable job. The emperor Tiberius, three Caesars before Nero, was once so displeased with a horoscope that he had the poor wretch who cast it bound hand and foot, weighted, and heaved into the Tiber.

It was a hazardous profession.

The tale is told of the court astrologer of King Louis XI of France (reigned 1461–83). The astrologer predicted—accurately, as it turned out—the death of a lady in whom the king had taken a more than strictly regal interest. When the lady died as foretold, the king, stricken with grief and rage and ready to blame the messenger for the message, summoned the astrologer into his presence. Beforehand, he had arranged a signal with his guards; when they got it, they would seize the luckless stargazer, sew him into a sack, and drop him into the Seine. When the man arrived before him, the king said: "You are right able, so God has showed unto us, to see into all of the future, foretelling it as it shall come to pass. Therefore I ask unto you, tell me how much time does remain to you yourself to live upon this earth?" Replied the astrologer: "In truth, Lord King, I know only this: The stars have spoken to me and said I shall remain on earth a living man until three days before the death of Your Majesty, which pray God will be a long time hence." Apparently King Louis shared this sentiment, for he never gave the order to chuck the man into the river, and instead is said to have treated him particularly well from then on.

Not all rulers and princes have taken the prophecies of their soothsayers so utterly to heart. According to Dion Cassius (who wrote in the third century A.D.):

> Many prodigies preceded the death of Vespasian [Roman emperor A.D. 69–79]: a comet was seen for a long time and the tomb of Augustus opened by itself. When the doctors reproved the Emperor for continuing to live as usual and going about state business after being stricken with a serious illness, he replied, "An Emperor has to die on his feet." Seeing some courtiers discussing the comet in an undertone, he laughed and said, "That hairy star has nothing to do with me; it is more of a threat to the King of Parthia, since he is hairy and I am bald."

This is in the same spirit as Hannibal's reply, more than 250 years earlier, to the king of Bithynia, when the sovereign became anxious to avoid a battle because of an unfavorable reading of entrails. Said the great Carthaginian warrior: "So you prefer the opinion of a sheep's liver to that of an old general?"

III Tidings

In the fourth century B.C. Aristotle described one comet that had a tail sixty degrees long, or across one third of the sky; the philosopher believed that its appearance signaled the decline of Sparta. To Ephorus, the same comet presaged the destruction of Helice and Bura, two cities in Achaea swallowed by the sea. Later, the historian Sozomen detailed the apparition of a comet shaped like a burning sword, which appeared above Constantinople in A.D. 400, when the city was threatened with betrayal at the hands of the traitor Gainas.

We shouldn't snicker. Comets have established a better track record than even Al Capone when it comes down to the business of thinning out the populace. Cometary appearances have either heralded or marked exactly these notable deaths:

Constantine (336)
Attila (453)
Emperor Valentinian III (455)
Emperor Maurice (602)
Louis the Debonair (837)
Emperor Louis II (875)
Boleslaw I, King of Poland (1024)
Henry I of France (1060)
King Harold of England (1066)
Pope Alexander III (1181)
Richard the Lion-Hearted (1198)
Philip Augustus (1223)

Emperor Frederick II (1250)
Giangaleazzo Visconti, Duke of Milan (1402)

Duke Giangaleazzo was a miserable tyrant, already ill abed when the comet appeared in the sky over Europe. Beholding the fatal messenger, the duke abandoned all hope:

> For our father revealed to us on his deathbed that, according to the astrologers, such a star was to appear for eight days at the time of our death. I give thanks to God for wishing my death to be announced to all men by this sign from heaven.

Resigned the duke may have been; humble, no. It was all the same to the comet, which snuffed him anyway. After all, that was the business of comets, as the historian Nicetas knew so well when he recorded the horrible details of the visit of the comet of 1182:

> After the Latins had been driven from Constantinople, an omen was seen of the rages and crimes to which [Emperor] Andronicus was about to abandon himself. A comet appeared in the sky; like a coiled snake, it sometimes stretched out and sometimes bent back on itself. Sometimes, to the horror of the onlookers, *it opened an enormous snout,* as though, greedy for human blood, it was about to drink its fill.

But let's have a little fresh air in here.

Bold King

In 1664, Alfonso VI, king of Portugal, received news of the appearance of one of the deadly visitors above his kingdom. The king was inside the palace, hard at work on state business, when the news was whispered in his ear. With a great bellow the king leapt to his feet, driving his courtiers from him with heavy blows as he galloped across the huge hall and headed to the door. Rounding a corner, he thundered off along a corridor, burst into a small armory, and snatched a brace of pistols. Thrusting the weapons into the hands of his astonished armorer, His Majesty bellowed at the man to load them and follow quickly.

Banging from the room, he pounded up four flights of stairs and staggered onto the lofty promenade, gasping for air. Regaining his breath, the king directed a volley of such vulgar abuse at the menacing comet that his retinue was frozen into silence. Then he grabbed the now-loaded pistols, brandished them at the comet, and, cursing wildly, ordered the wretched star from his realm, directing it toward Spain. Doubtless frightened by the raging king, the comet left. Why more kings did not adopt this simple, no-nonsense attitude to comets, the historians do not say.

Halley's comet has passed perihelion (the point in its path that is closest to the Sun) thirty-two times since 467 B.C., which is the first time an observation of the famous comet appears in the records of man. In 837, Halley's appeared during the reign of the French king Louis the Debonair, whose name appears on the list of comet victims above. A chronicler of the time, known only by the nickname the Astronomer, described it this way:

During the holy days of Easter, a phenomenon which is always ominous and a carrier of bad news appeared in the sky. As soon as the Emperor [Louis was also Holy Roman Emperor], who always paid great attention to such events, had noticed it, he allowed himself no rest. "A change of reign and the death of the prince are announced by this sign," he told me. He consulted the Bishops, who advised him to pray, build churches, and found monasteries.

Which he did, but he died three years later.

In 1066, while William the Conquerer was marshaling his army in Normandy to cross the English Channel and drag King Harold from his throne, the *Anglo-Saxon Chronicle* made this observation:

In this year King Harold came from York at Easter, which was after the mid-winter in which the King [Edward the Confessor] died. Then was seen over all England such a sign in the heavens as no man ever before saw; some men say it was the star Cometa, which some men call the haired star; and it first appeared on the eve of Litani-major, the eighth of the Kalends of May [April 24], and so shone all the seven nights.

The star was Halley's comet. The Bayeux Tapestry, the magnificent embroidery that constitutes the Norman record of William's great conquest, depicts King Harold sitting on his throne as the comet passes overhead. The king looks worried, as well he might. He was slain at Hastings, along with many of his nobility, when he met Duke William in battle. To this day the battle is commemorated in one of the great crowns of state worn by the English sovereigns. A blazing jewel set into the glittering mass represents the tail of the fateful comet that wrested the throne from the English and gave it to the Normans.

The Sword of the Turk

The year 1456 was a dismal one for Christendom. Three years before, the imperial city of Constantinople had fallen to the sword of Islam. The implacable armies of the Ottoman Turk faced Europe on Europe's own soil. The holy and famous church of Santa Sophia became a mosque where the praise of Allah rose to claim for the Saracen one of the glories of Christianity. Christians had been slaughtered or led into slavery in the hundreds of thousands. It seemed as if Christ had abandoned his people, and as if to mark their fate with a seal of flame, Halley's comet appeared in May of that dreadful year.

Observers of the day reckoned the comet's tail at a length of two celestial signs, or about a third of the sky. Disheartened and fearful, all Europe looked to Rome.

The pope, Calixtus III, recognizing the mortal danger signified by the comet, begged the princes of all Christian kingdoms to turn their arms against the Moslem invaders. As a further measure, His Holiness ordered the faithful to pray, and directed the bells to be rung at noon across the whole of the Christian dominion. The pealing bells of France, the German principalities, the Papal States of Italy, and the kingdoms of Spain—from the Baltic to the Aegean—all were to ring like voices of supplication, bearing up to heaven the prayer to the angels of God, the Angelus. The tradition of saying the Angelus at noon dates from this time of Christian peril, and has continued even to today in some parts of the world.

We should remember that dread of these heavenly apparitions was not the exclusive occupation of a credulous peasantry. The popes and potentates who took it all seriously were men of considerable education. Ambroise Paré was a famous surgeon and a man

29

with a solid reputation as a scientist. Here is a scholarly description wrenched from him by the comet of 1528:

> The comet was so horrible and so frightful and it produced such terror in the vulgar that some died of fear and others fell sick. It appeared to be of excessive length and was the color of blood. At the summit of it was seen the figure of a bent arm holding in its hand a great sword, as if about to strike. On both sides of the rays of this comet were seen a great number of axes, knives, and blood-colored swords among which were a great number of hideous human faces with beards and bristling hair.

If anyone benefited from all this vigorous terror, it was the Church. Always prepared to believe that God was ready to scrub the whole planetary experiment with one ill-tempered backhand, the citizenry of Christendom wisely resolved to pave their way into the next world with whatever they managed to put aside in this. And so when a comet appeared, such as that of 1528, demonstrating absolutely that God had finally had it with Earth, the prudent man shelled out. Endowments poured into monasteries to secure prayers for the souls of the benefactors. Just how the monastery was supposed to discharge its part of the deal after the Creator flattened the whole shebang, we are not told. Nor, presumably, were the donors encouraged to dwell upon this irony.

Rare Smilers

There were, of course, at least some observers mulling matters over and coming up with answers less sacred than sardonic. Erasmus of Rotterdam, writing about the same time as Paré, said:

> Would to God that wars had no other cause besides the bile of sovereigns, heated by some comet. A skillful physician would soon restore the blessings of peace with some prescription of rhubarb.

Madame de Sévigné wrote this in 1681 in a letter to the comte de Bussy:

> We now have a comet which is very extended, with the finest tail imaginable. All the great ones are alarmed, believing that Heaven is preoccupied with their deaths and warning them by means of the comet. They say that when Cardinal Mazarin despaired of his doctors, his courtiers felt obliged to honor his agony with a miracle and told him that a great comet had appeared which frightened them. He had enough self-control to laugh at them, telling them ironically that the comet did him much honor. He was quite right too; human pride does itself too much honor in believing that there are big upheavals among the heavenly bodies when one is about to die.

Alas, the limpid voice of Madame de Sévigné was rather isolated in the general babble that rose from the court of the Sun King. This is from the *Chroniques de L'Oeil-de-Boeuf;* the comet is that of 1680:

> All the telescopes are trained at the sky; our learned Fellows of the Academy of Science are

busy day and night with a comet the like of which has not been seen in modern times. There is great terror in the town; the timorous see it as a sign of a new deluge since, as they say, water is always announced by fire; this does not seem to me to be a convincing argument, unless M. Cassini [the court astronomers]— should take the trouble to confirm it. While the fearful are making their wills and leaving their goods to the monks in expectation of the end of the world, there is high agitation at court on the question as to whether this wandering star has come to announce the death of some great personage, just as they say it announced the death of the Roman dictator [Caesar]. Yesterday some of the bolder spirits mocked at this opinion; Louis XIV's brother, who seems to be afraid of becoming a Caesar all of a sudden, drily exclaimed, "Well, gentlemen, it is easy for you to talk about it; you are not princes."

The comet so raptly tracked across the sky by the courtiers of the Sun King held spellbound half the human race. Catholic and Protestant, Turk and Jew, all were united in an ecumenism of fearful speculation as the burning mark inched slowly through the canopy of stars. And humans weren't the only creatures startled by the apparition. The world of poultry seems to have joined in the general alarm. Camille Flammarion, the nineteenth-century astronomer and popularizer of science, rooting around in the archives of the Bibliothèque Nationale in Paris, found a print from 1680 titled *The extraordinary miracle which occurred when a hen in Rome laid an egg on which a picture of the comet was engraved.* The engraving bore a legend testifying that the truth of the affair was "certified by the Pope and the Queen of Sweden." The defense rests.

The twentieth-century reader, accustomed to such

features of modern life as the entirely plausible threat of nuclear obliteration, might find it hard to rustle up much sympathy for the cometary terrors of his ancestors. So a comet rips across the sky and tears the life out of a few despots on the way past. So what? Well, there were once grimmer scenarios than that.

Fried by the Sun

William Whiston was an Englishman, an astronomer, and a theologian . . . all at the same time. This would probably be too much for one man to handle even today. In the seventeenth century, frail man was simply unequal to such an appalling confluence, as becomes apparent from a perusal of Whiston's *New Theory of the Earth*, published in 1696.

Whiston attempted to reconcile advances in geology with the chronology of Earth set down in the Book of Genesis. This he did by interpolating a comet into the fray. Edmund Halley had computed the orbit of the 1680 comet, assigning it an orbital period of 575 years. Whiston seized this figure and, clutching it tightly, leafed diligently back through his stack of tomes until he found a date for the Flood that matched a visit by the comet. Now the astrotheologian (theo-astronomer?) could really get cracking.

God knew all along, according to Whiston, that man was going to be a sinful creature. What's more, He knew that eventually our sins would get to be too much for even Him to take, and that He would straighten us out in a major way. This did not mean calling us in for a chat; it meant curtains. It happened as follows.

The date was either Friday, November 28, 2349 B.C., or Tuesday, December 2, 2926 B.C. On one of those dates—Whiston lets you take your pick—the comet went scowling past Earth a scant 10,000 miles above our heads. It crossed the meridian of Peking, which is where Noah was living at the time. Whiston does not say what Noah was doing in China, although he may have been hiring cheap labor for work on the ark. At any rate, the close passage of the comet raised monstrous tides and broke open the Taurus Mountains of Armenia, letting all the water that normally sloshes about beneath the Earth's surface spew up into daylight. As if this were not enough (it wasn't), the comet's tail slapped against the terrestrial atmosphere and started the rain pouring down. This lasted for forty days . . . no surprises there. When it was all over, according to Whiston, the water was five miles deep. We are not told who took the sounding, but it may have been the clever chap who helped Whiston figure out what would happen next.

At some time in the future, the comet will pelt past again, this time so close that it will jolt Earth's orbit nearer the Sun. Too near. We will all fry. There will follow a period of 1,000 years when the earth will be ruled by saints, although why anyone would want to exercise dominion over a planetful of ashes is unexplained.

At the end of this millennium of charred, if righteous, government, the comet will return to strike poor, battered Earth, turning our planet into a comet too! The saints will go marching home. No one can say Whiston didn't know how to tell a story.

Duck!

In 1759 the reputable French astronomer Joseph Jérôme Le Français de Lalande published a volume entitled *Reflections on the Nature of Comets*. In his book Lalande discussed a few comets whose orbits might, in some circumstances, bring them closer to

Earth than normal. This was seized upon by educated people and presented as prophecy that a comet was headed straight for Earth. From the nobility this apprehension seeped down to affect all ranks of French society. At every level, the citizenry began to neglect the daily affairs that keep communal life going—bakers shut down their ovens; farmers abandoned their fields. The kingdom was grinding to an anxious halt as France laid down its tools to await the impact that would reduce creation to a smoking lump of ash. Finally the king intervened. He ordered Lalande to publish another document, this one explaining what he meant by the first. Only a direct and unequivocal explanation by the scientist himself—at the clear behest of the royal power—was enough to quiet the fears of the populace and get France back to work.

Even as recently as the last century, predictions of a cometary collision terrified people. This time, the predictions seemed to be so securely founded on the solid and inexorable principles of science that there could be no doubt of their veracity. The comet was going to smash into us, said the little figures scribbled in the astronomers' notebooks, and that was that.

This prediction concerned the return of Biela's comet in 1832. Computing the comet's course, a French astronomer concluded that the comet would pass through the plane of Earth's orbit before midnight on October 29, 1832. The comet's center would come within 18,000 miles of Earth. However, the comet's head was 21,000 miles wide. Therefore part of the comet would be in the same space as part of Earth; thus, *ka-powey!* The newspapers, resolute in their duty to publish the smallest detail of anything so wonderfully replete with dread, laid it all out for the public, who resignedly buckled up for their last hurrah.

But wait a minute. Someone, somewhere, had forgotten something. Oh, they knew where the comet was going to be at midnight on October 29, 1832, all right: it was going to be at a certain point on the orbit of Earth. But where was Earth going to be? After all, Earth's orbit is 738,030,000 miles long, which gives it a lot of places to be on any one day. The physicist François Arago sat down to figure it all out. He found that when the comet was tearing through that one spot on the terrestrial orbit, Earth would be loafing along at its usual speed of 84,000 miles an hour, *some 60 million miles away from where the planet was intersecting its path*. This was a good month's hike for Earth, which wouldn't get close enough to the comet to feel the wind on its cheek.

Without a Button to Push

That these computations were available to science at all was due to the slogging of two seventeenth-century collaborators, Sir Isaac Newton and Edmund Halley. Halley, working with Newton's theory, hammered together an extremely elaborate construction of mathematical calculations, determining that the comet of 1682 would come around again in either late 1758 or early 1759. Part of Halley's task would have been to figure out just how to figure it out. In other words, the mathematical formulae that today make such predictions simple did not exist for Halley. Later, Alexis Clairaut refined the algebra of the calculations. Even then it took two computers—humans, not Univacs—six months of toil to work out the formulae developed by Clairaut. Clairaut's work enabled him to predict that Halley's comet would be delayed 100 days by the pull of Saturn, and another 518 days by Jupiter. He estimated mid-April of 1759 as the time of perihelion passage. In fact, Halley's passed perihelion on March 12—

confirming an astonishingly close piece of computation for the tools the Frenchman possessed.

To understand properly how remarkable the achievement of those earlier mathematicians and astronomers was, remember that at the time of Halley's prediction the outermost known planet was Saturn. Saturn patrolled the very edges of man's universe at a distance from the Sun of 900 million miles. Compare this to the distance from the Sun of the object whose return Halley was predicting, 3.2 billion miles, and you will see how daring a feat was the forecast.

And so into our own century, where dwell the smooth, silicon rabbis of technology. The twentieth century: Apollo, Challenger, moonshot, Canaveral, Columbia, booster, Saturn, module, lunar. The words dance across our consciousness like the calibrated poetry of scientific worship. We are inventing a language to carry into space, and the language is crafted to enchant us with the wonder of man's dominion.

Mulling Over Mars

But our century is already old. When it was still young, men were not compelled to await the results of probes and computer analyses before postulating what lay out there. Not actually having visited the place, we could think of Mars what we liked. When Halley's comet came charging across the skies of the Sunward planets on its last visit in 1910, Americans were entranced not only by the advent of the famous comet, but also by the lively theories of Professor Percival Lowell.

Professor Lowell believed that the apparent canals of Mars were real canals, and that they were lined with vegetation watered by the melting Martian ice caps. He believed that Mars was running out of water, and that its population faced extinction. Arriving in Plymouth for a lecture tour of Britain, Lowell was asked by reporters to explain his vision of the Martians (the quotation is from the *New York Times* for March 26, 1910):

They are not human beings. The comparative size and the conditions, every known factor in the planet point to something very different. They are intelligent organisms, but not in the least like man.

The reporters asked Lowell if he thought that Earth faced the same fate as Mars. His answer:

All planets must come to an end some day. The world is larger than Mars and its water may last longer. Mars is now dying of lack of water; Venus, with one side always towards the Sun and the other in perpetual cold and night, is dying of paralysis. What the end of the world will be who dare foretell?

He was a forbidding fellow, Professor Lowell. Perhaps it was his heavy-handedness that sparked the *New York Times* editorial on the following day, surely written by a man with his tongue securely in his cheek:

Presumably the Martians are not mammals, and the theory that they are large and highly developed ants may now come into new favor. It is not a new theory and it has always seemed reasonable, comparatively speaking. Pure reason, of course, has very little to do with speculations regarding life on Mars.

The ant is an industrious digger, with a sense of form, and appreciable architectural ability. If the Martians are ants, however, it is to be hoped that they will always stay on Mars, and not attempt to emigrate to the Earth when the conditions of existence in their own abode become unbearable. Why should they not be able to adapt themselves to any conditions? The ants we have always with us can live with very little water or air. A colony of ants will inhabit the remote recesses of a sugar bowl and thrive wonderfully. We may hope that Professor Lowell will now take up the ant theory and work it out.

He seems to have a fresh rival in the field of imaginative astronomy. Master Sidis, Harvard's fourth dimension expert, has invented a new airship, with a radium engine, which will carry him to Venus in twenty minutes. But it could carry him to Mars in less time, and he can learn more about the canals and the inhabitants in twenty minutes than Professor Lowell can learn by the aid of the telescope in a lifetime. The radium engine would be rather expensive at present, but Sidis is young and the cost of living will surely be decreased in a few years.

Of course.

Halley's 1910

But fascinating as interplanetary travel and the physiology of the Martians must have been, most Earthlings reserved their anxieties for the approach of Halley's comet. It was business as usual, with letters flooding the offices of government and science from people wondering about the best way to prepare themselves for the coming impact. So annoyed was Sir Robert Ball, the Cambridge astronomer, that he wrote to the *Times* of London enclosing a form letter which he begged the newspaper to publish.

> *My Dear _____,*
>
> *A rhinoceros in full charge would not fear collision with a cobweb! And the Earth need not fear collision with a comet.*
>
> *In 1861 we passed through the tail of a comet and no one knew anything about it at the time.*
>
> *For a hundred million years life has been continuous on this earth, though we have been visited by at least five comets every year. If comets could ever have done the earth any harm they would have done it long ago and you and I would not be discussing comets or anything else.*
>
> *I hope this letter will give you the assurance you want. So far as I can learn we may be in the tail of Halley's about May 12: and I sincerely hope we shall.*
>
> *I think Sir John Herschel said somewhere that the whole comet could be squeezed into a portmanteau.*
>
> *Yours truly,*
Robert S. Ball

Sir Robert's sangfroid might be all very well at Cambridge. At Constantinople it would have seemed a decidedly alien attribute. This is from the *Times* of London:

> Constantinople is a city where every beast of burden and most children wear charms to guard them against the evil eye, where astrologers still ply

a gainful trade, and eclipses are watched with terror by a goodly portion of the people. Small wonder that Halley's Comet was regarded with awe and that a tale of how certain Frankish astronomers had foretold the destruction of the Earth found credence among many, and aroused an obscure uneasiness among more of the Christians and Moslems of the Turkish capital. On Tuesday night, so men averred, an aerolite [meteorite] fell in the Yildiz Park, but this occurrence, which had surely been fraught with the most portentous and awful significance had Abdul Hamid [the sultan] been still there, rather relieved the old ladies of Stambul. "Now," they said, "the comet, please God, will be less dangerous since its tail has fallen off."

Yesterday was a day of rain and thunder, and businessmen remarked that their clerks showed an uneasy nervousness at each peal. With the fall of night the timid and the curious came out upon their roofs, and well nigh a hundred thousand souls passed the night there, some to watch the comet, others to encourage the fearful, others out of a natural desire to meet the terrors of the last day in the company of friends and neighbours. Some prayed, others sought to cheer or deride their fellows with songs and music, and many, when the dawn had come and the natural order of the Earth and the Heavens were seen to have suffered no interruption, cheered and clapped their hands lustily, to the annoyance of such who wished to sleep.

Meanwhile, the municipal authorities, undeterred by celestial signs and portents, at last fulfilled their threat and marshalled their levies against the famous street dogs of Constantinople. All through the night in more than half the quarters of Pera and Stambul, detachments of police-man and sweepers, armed with lassoes and huge wooden tongs and followed by an impressive array of dust carts, raided the unsuspecting and familiar beasts. Some they lassoed, others they gripped with their tongs and hoisted for all their squeals and yelpings into the dust carts. A few were spared at the intercession of soft-hearted inhabitants, who were, however, compelled to go bail for their protégés and to swear to provide them with collars bearing their names and addresses and to pay the dog-tax when the municipality should so decree. The captives were driven away—where, it is uncertain. Some aver that they will share the fate of the other "reactionaries" and be marooned on more or less desolate islands, others that they will be converted into gloves, others, again, that they will be pensioned off and live out their lives in concentration camps on municipal rations. Allah bilis!

What with their protests against the treatment of the dogs and their alarms on the account of the comet-gazers, it is not surprising that many of the good citizens of Constantinople passed a very wakeful night.

Back in the United States, the Yerkes Observatory (in Williams Bay, Wisconsin) reported that spectra of the comet's tail revealed the presence of the poisonous gas cyanogen. The *New York Times* reporter covering the story felt he should add that "a grain of its [cyanogen's] potassium salt touched to the tongue is sufficient to cause instant death." Just to make sure everyone got the point, the *Times* also reported Professor Camille Flammarion as holding "the opinion that the cyanogen gas would impregnate the atmosphere and possibly snuff out all life on the planet."

The point was taken. When Earth passed through the comet's tail, miners in the towns of Leadville and Cripple Creek, Colorado, spent the dangerous hours underground to avoid the deadly fumes. A couple of thousand miles away in the anthracite coal mining region of Pennsylvania, other miners adopted radically different tactics. Certain that the world was going to shudder to an end when it passed through the tail of pollution spread by the passage of Halley's comet, these miners *refused* to go underground, preferring to meet their Maker in the open air. The mineowners might have avoided these interruptions of business by availing themselves of the remedy supplied by one enterprising American. He offered anticomet pills, a bargain at a dollar a box, and guaranteed that they would ward off the noxious gases of any passing star.

From every point on the globe came reports of cometary influence: people dancing in the streets; frightened worshipers jamming the churches; men and women going insane; suicides. In Switzerland, groups of the passionately curious headed skyward in balloons and on funiculars to get a better view. In New Jersey, one woman chose the propitious apparition as a sign that she should return to the still-searching husband she had abandoned a few weeks earlier. So insidious was the lure of Halley's that it even managed to lodge its appeal in one of America's most bemused and ironic minds, that of Mark Twain.

The great humorist was born in 1835, when Halley's paid one of its regular calls on the Sun. In 1909, the year before the famous comet was due for another visit, Twain wrote these lines:

I came in with Halley's comet in 1835. It is coming again next year, and I expect to go out with it. It will be the greatest disappointment of my life if I

43

don't go out with Halley's comet. The Almighty has said, no doubt: "Now here are these two unaccountable freaks; they came in together, they must go out together."

Oh, I am looking forward to that.

Sure enough, Mark Twain died in 1910.

CHAPTER 2

Out of the Frozen Cloud

For countless millennia—no one knows how long—Halley's comet has slid along the path of its awesome ellipse, whispering through interplanetary space at a velocity of 100 miles per second. In 1910 it paid its latest visit and was greeted by the special lunacy that Earthlings reserve for their favorite invaders. Many of these antics are humorous today, but they were not always an unqualified hoot for the people involved.

In Oklahoma, for example, a group of zealots called the Sacred Followers believed that Halley's was a signal of divine displeasure and, more than that, due notice that the Creator planned to undo a little of His handiwork in a brutal fashion. Being lunatics, they felt that there was something they could do to deter the Lord, even though He seemed to have made up His mind. Seizing a young virgin prisoner, the Sacred Followers prepared to murder her as a sacrifice to the wrathful deity. What possible use God could have for the corpse of a young woman, we are not told. Fortunately, a posse of Oklahoma sheriffs' deputies was able to intervene in time, and the execution was forestalled.

Never assume that comet fever is an affliction that eases with the passage of time, or that the more scientifically literate people of today will regard the return of Halley's entirely with dispassion. They were not scientific innocents who crowded atop the hotels of New York in 1910 to party the night away in honor of the legendary comet. Certainly they were not dragging virgins aloft to hurl to the pavements below, but there was plenty of speculation of a decidedly spirited nature. And a scant decade before the 1910 appearance of Halley's, the *Fortnightly Review* of London was carrying an article on "The Menacing Comet":

In view of the interest awakened in the subject of a possible rencontre between the Earth and Tempel's Comet, it may be well to consider how far Professor Falb is justified in making his much talked-of prediction, namely, that on the 13th Nov., 1899, the comet will come into collision with our planet. In a lecture delivered subsequently in Berlin, he explained that his prophecy was to be regarded merely as an announcement of the fact that on the date mentioned, between the hours of two and three in the morning, Tempel's Comet will be due at a point on the Earth's annual path, where the Earth also will arrive at the same time, hence the predicted collision. He disclaims all thought of exciting alarm in the public breast; nor would he wish to . . . tell the world exactly what will or will not happen. But of one thing he feels pretty sure, namely, that the material of which the comet is composed is so light that the inhabitants of this nether world may confidently expect to come out of the tourney scatheless; unless, indeed, the comet's atmosphere should prove to be of a noxious character and overwhelm us with a torrent of carbonic gas.

Sadly, the public breast was not nearly as sanguine about the onrushing comet as was the untroubled Professor Falb. It was the practice of the populace to receive any news of comet collisions in their near future with general disorder in the streets. This may be pitiful, incurious to a fault, and unscientific. But there you are; what is to be done with humans? You cannot simply rope them into place and speak severely to them. Not when they are getting this from the *Lettres Cosmologiques:*

Reflecting on the laws of gravity, it will be readily perceived that a comet's approach to the Earth might cause the most woeful events; bring back the universal deluge, or make the Earth perish in a deluge of fire! Shatter it into small dust! Or at least turn it from its orbit; drive away its moon, or still worse, drive the Earth itself outside the orbit of Saturn, and inflict upon us a winter several centuries long, which neither men nor animals would be able to bear!

Too true! The few sodden months we get right now are enough for most of us. But the observer who quoted the above in his survey for the *Fortnightly Review* was a man of strict balance. That was the stick. Here's the carrot:

Maupertius, however *(Lettre sur la Comète),* surveys these things through a less distempered vision. After enumerating the various ills with which . . . this lower world would be afflicted by the too near approach of a comet, he sees bright possibilities looming on the horizon on the visit to our planet of one of these wandering stars. "Might there not be changes," he asks, "of the seasons into perpetual spring? Our visitor might become a permanent guest, and accompany the moon in her revolutions; or give to us a ring like that of Saturn." And as to the dread of a collision, he goes on to say that however dangerous might be the shock of a comet, it might be so slight that it would only do damage to that part of the Earth where it actually struck; perhaps even we might cry quits [accept it] if while one kingdom was devastated, the rest of the Earth were to enjoy the rarities which a body coming from so far might bring to it. Perhaps we should be very much surprised to find that the debris of these masses which we dread were formed of *gold and diamonds!* And who would be the most astonished, we or the comet-dwellers who would be cast on the Earth? What strange beings each would find the other!

All of this for a lump of ice.

Sandbanks and Snowballs

A comet *looks* like a pretty substantial object as it blazes across the sky. The head of a comet, which astronomers call the *coma,* is usually denser and brighter than the tail, and occupies a volume of space many times greater than that occupied by Earth. In addition to this great bulge, there is a cloud of hydrogen wrapped around the coma more voluminous even than the Sun. And yet there is nothing to it! A comet is all show. If it were otherwise, if a comet represented even a fraction of the mass it looks to be, then the passage of a comet would have an effect on the planets it brushes past in space. But it is just the other way around. It is the planets that alter, or *perturb,* the orbit of the comet, overwhelming the insignificant mass of this little bird of space with the latent power of their forceful presence. In 1910 an American astronomical expedition to Hawaii planned to place itself so that members would be able to observe the nucleus of the comet in silhouette and determine its size once and for all. They saw nothing. Yet they were using telescopes that would have picked out a solid mass no larger than a thousandth the size of the Moon. Out of this failure grew the theory of cometary construction known as the flying sandbank.

The flying sandbank theory described the comet as nothing more substantial than a loose swarm of separate particles. Each particle was thought to be orbiting the Sun more or less on its own, but traveling with the rest of the flying sandbank because of the gravitation exercised by the whole. This theory remained popular right into the 1970s, probably because it explains so well some of the more persistent puzzlers about comets, like how such splashy, blustering fellows can be so hard to see when you look closely. Even today the flying sandbank has a few important adherents. But for the most part it has been replaced by the dirty snowball.

The dirty snowball was conceived, packed together, and heaved into the small but contentious world of comet science by Fred Whipple, a former director of the Harvard-Smithsonian Astrophysical Observatory in Cambridge, Massachusetts, and a pioneer of the American space program. Whipple decided that comets did have something solid to them, although he agreed that it was not very much when you considered the observable size of the objects. According to the Whipple model, Halley's comet is a big, lumpy, roughly spherical ball weighing in at about 65 billion tons. It has a diameter of about three miles and spins on its axis once every 10.3 hours. About half of the comet is made of dust and stone, and the rest is watery ice (40 percent) and other volatile materials (10 percent).

The Pressure of Light

The tail of a comet always points away from the Sun, whether the comet is approaching perihelion (its closest point to the Sun) or leaving it. A number of bizarre theories have been advanced over the ages to account for this phenomenon. Sir Isaac Newton believed that the Sun was actually nourished—given new life and energy—by the head of a comet. The comet's tail, on the other hand, replenished the atmosphere of Earth and the other planets with all the vital juices we need down here. The comet's tail, reasoned Newton, pointed away from the Sun because sunlight heated particles in the "aether" surrounding the comet and drove these particles away against the solar gravity in the same way that smoke is forced up a chimney by heat. Newton was wrong, of course, for space is too thin to sustain heat the way our atmosphere does; if it were otherwise, a comet would simply melt away on its first pass. But Newton was on the right track, believing that some solar force was at work on the comet. Johannes Kepler made the necessary adjustment. Not heat, said Kepler, but light. And according to today's theory, Kepler was correct.

Here in the thick, homey atmosphere that lies upon our planet like a snug blanket, it is not so much the force of light that strikes us when we walk outdoors as it is the force of great heat. On Earth, the force we feel from the Sun is heat, but in the thinness of space it would be light. If you drop some very light particles down a glass tube and then shine a light at them as they slowly sink, you may be surprised at how violently they jump and smack against the sides of the tube. It is exactly this same force that drives particles off the comet's core, or *nucleus,* and into the brilliant tail that appears to burn behind the comet like a plume of flame.

The tail is really a long stream (in the case of Halley's, many millions of miles long) composed entirely of these minute bits of comet, most of them only a few thousandths of a millimeter in diameter. The same force that dislodges these particles from the nucleus and drives them off into the tail also lights them up. What we see from Earth is simply the illuminated dust, shining with reflected sunlight. While the particles are relatively densely packed, the tail stands out brightly in the dark sky. But the particles do not long remain together in this tight formation. The radiant light continues to propel them away from the Sun, and as they fly farther and farther away the distances between the tiny particles increase. Finally, the little dots of light are so far apart that their reflections, unsupported by the nearness of their fellows, are too weak to detect. Thus it is that the tail appears to dwindle away like a spent flame. So strong is the force of sunlight on these specks of matter that some of them are driven clean out of the Solar System.

This force, the force of sunlight in nonatmospheric conditions, is one of the forces working to pin the tail on the comet. But it is not the only one. There is also the solar wind.

The Solar Stream

Every single day, the Sun bellows forth into space a massive volume of its substance, at the rate of about a million tons in twenty-four hours. The substance blown away from the solar surface in every direction is called *plasma*. This plasma is electrified gas, gas in which the constituent molecules carry electrical charges imparted to them by the tremendous magnetism of the Sun. These electrified molecules, or fragments of molecules, are called *ions*. It is these ions of electrified gas pouring out of the Sun into space that constitute the plasma stream of the solar wind.

What happens when a comet encounters the plasma flow of the solar wind roaring off the Sun at 900,000 miles an hour?

The nucleus of a comet is surrounded by a huge coma, the vast, luminescent cloud that, with the nucleus, makes up the comet's head. The coma itself is wrapped in an envelope of hydrogen gas. In the case of Halley's, this whole is huge, and it balloons along through space, surrounding the rather miserable little ball of ice and rock—the nucleus—at the root of the spectacular celestial contraption. When the fierce plasma torrent of the solar wind boils into this gassy head, the electrical charges of the Sun-born ions in turn ionize, or charge, the molecules in the bulging head of the comet, and more plasma is created. The rushing solar wind sweeps this fresh plasma into its train as it blasts off past the comet, and the new plasma is woven into the bright tail of dust already there, adding its own light.

The light created by the plasma is not the same as the light of the particle tail. The plasma fluoresces; that is, the atoms or molecules shine with their own luminescence, releasing their energy into the wind that bears them along. Russian astrophysicists have constructed a device at one of their laboratories in Moscow that fires ionized hydrogen gas at speeds much greater than those of the solar wind. The "comet" they bombard with this ruthless punishment is a ball of wax. Vapors pour from around the waxen ball and form a tail; both the tail and the impressive head surrounding the ball glow like a real comet.

Scientists believe that the solar wind may be behind the disturbances that result in aurora borealis, the beautiful and enchanting northern lights that spill across the sky in northern latitudes on certain nights. There is a great bulge of magnetism surrounding Earth, in some places as thick as 60,000 miles. This giant pad, and the thickness of Earth's atmosphere, help our planet to shoulder its way through the solar wind untroubled. But once in a while the Sun gives forth a particularly breathy belch, or there is some temporary weakness in the magnetism, and the solar wind pours through Earth's magnetic shield and dumps a load of glowing plasma—the aurora borealis—into our atmosphere.

Naming Your Comet

There are only about 700 comets listed in the official catalogue kept at the Harvard-Smithsonian Center for Astrophysics in Cambridge. This is the place to report any new comet, and have it duly baptized with your own name. Just fire off a telegram to TWX-710-320-6842 ASTROGRAM CAM, or write to:

Central Bureau for Astronomical Telegrams
Harvard-Smithsonian Center for Astrophysics
60 Garden Street
Cambridge, Massachusetts 02138

Make sure you also write *after* you have sent the telegram, of course. Your communication should specify when and where you made your sighting; the position, brightness, and appearance of the comet sighted;

and how fast it appeared to be moving against the backdrop of stars. If you think, gazing out there some night, that you have discovered another comet, chances are good that it's already listed. Or that you are watching some wonderful abnormality, the true location of which is your own telescope. But if what you see persists in being a comet, it will fall into one of these entirely arbitrary divisions, which please feel free to throw out and replace with more sensible ones of your own:

The Trillionaires—So far about 100 of these have been logged, long-distance travelers who have journeyed bravely anywhere from 3 trillion to 6 trillion miles before they come streaming in to whip around the Sun and loop off again on their inconceivable orbits. The most famous of these was Kohoutek, which stirred up a great deal of interest and gloomy speculation when it appeared in 1973.

The Millionaires—There are about 500 of these. Their orbits can take anywhere up to 2 million years to complete, although most of them have shorter orbits. The comet West, which appeared in 1976 on a 16,000-year orbit, is one of these.

Halley's Comet Comets—There are fewer than twenty comets with orbital periods of between twenty and two hundred years. These comets usually have orbits within, or at least not very far outside, the orbit of Pluto, the last planet in the Solar System.

The Homebodies—These are the space kids, so attached to the Sun that their orbits fetch them home every five to ten years. There are about 100 homebodies right now, but it is comets of this class that are most vulnerable to being perturbed right out of the Solar System by Saturn or Jupiter.

Encke—Encke is all alone, the only comet with an orbital period of only 3.3 years. Encke does not venture far enough from the Sun to be jostled off into space by the bullies who lie in wait beyond Mars.

The Great Cloud

According to the *Great Cloud* theory most favored by astronomers at present, all of these comets come from the same place, an enthralling sphere that surrounds the Solar System at a distance from the Sun of between one-half and a full light-year out. This sphere is composed of many billions of comets, and has existed much as it is now from the earliest days of our Solar System. There they exist, and there most of them will likely remain. All of them would remain there, of course, pursuing the leisurely life of a comet at home, ambling through space at a snail's pace in crowded company, were it not for the fact that the universe is a very violent place. After all, we are not alone here.

There are some 200 billion stars in the Milky Way, of which the Sun is only one. As the Sun drags us and our sister planets along on the endless whirling voyage around the center of the Galaxy, the spiral arm in which we travel shifts and settles, losing some stars, gaining others, as its thousand million suns wheel in and out of the gas and dust and freezing distances of interstellar space. Drawn along with the Sun on this perilous voyage are the comets, a great, jostling cloud of lumps of dirty ice. The distance from one side of the spherical shell that holds the comets, in across the Solar System, past the Sun, and out the other side, is 12 trillion miles. A 12-trillion-mile-wide target is pretty hard to miss, folks, even in space. And so get hit it does. Every 10 million years or so some voyaging star trekking through the spiral arm hurtles in upon the clubby band of comets and—*ka-pow!*—a few more members are suddenly missing from their places at the bar.

The star that sent Kohoutek plunging in toward the Sun probably cut him out of the pack 8 million years ago, and he has been on his way ever since. This doesn't mean that it will be another 16 million years before we see Kohoutek again. Comets often lose energy as they run past the planets, and Kohoutek was one of those. On the other hand, a comet may sometimes gain momentum from planets, accelerating so much that it tears back out to the sphere from which it came . . . and keeps right on going. As a general rule of thumb, if a comet crosses a planet's path in front of the moving planet, the comet will lose energy. If the comet passes the planet's path after the planet has sped by, then the comet gains speed, often enough to give it the heave-ho out of the Solar System forever.

This is not just information that is fun to know. The astronautical engineers and astrophysicists who plot the trajectories of some of our space probes have already used this planetary booster effect to alter the direction of a spacecraft's travel and shoot it off on a new course, at a different speed. Looking ahead to the days when interstellar travel is possible, physicists believe that spaceship commanders will use stars the same way in the future, either for enormous boosts of free energy or for a quick tramp on the interstellar brakes.

While all of the planets can and do affect the orbits of comets and help to determine their arrival time at perihelion, the main players are Jupiter and Saturn. In *The Comet Is Coming,* the British science writer Nigel Calder cites the effect of Jupiter on the comet Brooks 2 in 1886. Speeding in toward the Sun, the comet actually overtook the giant planet as it charged along through space. At its closest, the comet closed to within 60,000 miles of Jupiter, finally swinging around in front of the planet. Brooks 2 lost so much energy from this maneuver that its orbit was reduced *from twenty-nine years to seven years*.

The Age of Comets

How did all the comets get up there in the first place? No one knows. Fortunately, this does not mean that no one will supply an answer. Speculation is the fuel and the fiber of astronomy.

Four and a half billion years ago, when the universe itself was already 15 billion years old, our own solar system boomed into being out of the collapsing mass of dust and gas that had been drawing together in its own quiet corner of the Galaxy for 100 million years. Like all new stars, the proto-Sun at the center of this seething mess, unable to resist the gravity of its mass, finally collapsed into a tight ball, its density driving up the heat of the gas within until a thermonuclear reaction began and the Sun exploded in a gale of flaming material. This is just the standard star birth. All new suns throw off a raging wind of burning gas at first. It is their way of saying hello.

Closest to this furnace, only stone could survive, and so the stony planets, Mercury, Earth, and Mars, evolved. Next came the two largest planets, Jupiter and Saturn, which collected most of the spare hydrogen flying around. Beyond were the icy planets, farthest from the Sun. But as it turned out, there was still plenty of material left over when Uranus, Neptune, and Pluto were through building. It was these little odds and ends just beyond the gravitational reach of the chilly planets that eventually marshaled themselves into the vague and shuffling adherents of the great sphere. There would at that time have been many more comets than there are now, and Earth and all the other planets would have taken a constant drubbing from them. This was the Age of Comets, either an evolutionary nuisance to be endured, or an all-nourishing rain of valuable elements from our stellar womb. Take your pick.

Tracking the Comets

The theory of the cometary sphere, that is, a vast cloud of comets, originated only in 1950, and grew out of the work of Jan Oort, the paramount Dutch astronomer of his day and director of the famous observatory at Leiden. Oort came up with the comet cloud after extrapolating comet paths from the curve of a comet's route around the Sun. This is tricky work. Say that a comet comes from far beyond the Solar System, instead of from the very edge of it where the comet cloud is supposed to be. Its path around the Sun will not necessarily reflect this, for a planet may have dragged it down to a much slower speed and tighter orbit, perturbing it into an orbit forever *within* the Solar System. On the other hand, a comet originating within the Solar System could receive one of those vigorous shoves from Jupiter or Saturn, so accelerating it that its path around the Sun looks like the path of a comet coming from *beyond* the Solar System. Such a path is called a *hyperbola,* simply an open-ended curve. A comet originating *within* the Solar System ought to describe an *ellipse*—the same kind of flattened circle as that made by orbiting planets, except that a comet's orbit ought to be more flattened, or eccentric. But it's not just those two surly thugs the other side of Mars who can perturb a comet's orbital path. The comets can do it themselves.

While Jan Oort was scratching his head at Leiden, Fred Whipple was tinkering over his snowball theory at Harvard. One of the effects of his ice-and-rock composition, reasoned Whipple, would be a steady decline in the cometary mass as particles were swept away from the nucleus by the radiant light and the streaming solar wind. This jetting ribbon of material would, said Whipple, constitute a sort of rocket, the effect of which would be either to speed up or to slow down the progress of the comet, depending on the way the comet was spinning on its axis. When Oort factored this effect into his calculations, he found that *all* comets described ellipses, at least originally, and so were citizens of the Solar System.

Oort found that the inner edge of the belt of comets began at about 20,000 astronomical units from the Sun. (An astronomical unit [AU] is one Earth–Sun distance, roughly 93 million miles.) This same work also led Oort to determine the extent of the Solar System, and he found that it ended at 200,000 AU, or almost 19 trillion miles from the Sun. Beyond 200,000 AU the Sun's hold on orbiting bodies is too weak to prevent them from being torn from their places by another star. Otherwise the perturbation of a passing star would have the opposite effect, namely, to start the object falling in across the Solar System toward the Sun, and the watchful eyes of Earth.

The Sphere Within the Sphere

There is another source of comets to choose from. Dr. Jack Hills, a theoretical astrophysicist at New

Mexico's Los Alamos National Laboratory, believes that another cloud lies between the Oort cloud and Earth. In fact, he feels that this second cloud is larger, and that the Oort cloud is really nothing more than an outer shell or rim of the much more massive cloud within. What is more, comets of this inner belt, being more numerous and closer to the gravity of the Sun, are more tightly organized, less loosely defined, than comets in the outer region, and thus are less susceptible to being perturbed by the mere proximity of a passing star.

This inner cloud begins at about 3,000 AU from the Sun, 300 billion miles out. According to the Hills model, every 500 million years this inner cloud takes a direct slug in the chops from some monster the size of the Sun. This results in a comet shower lasting 700,000 years, and supplies us with 10,000 comets for each of those years. That accounts for 7 billion comets banged out of the inner cloud every 500 million years since the birth of the Solar System. Simple arithmetic yields nine of these batterings, accounting for 63 billion comets. How many can there *be* out there? Ten trillion, says Hills.

If even a fraction of the comets dumped into the inner Solar System over all these millions of years were still circling the Sun, any nighttime skyward glance would reveal a starscape scored with innumerable slashes of light. That it is not so indicates that most comets do not remain with us for long. Where do they go? Jupiter and Saturn, steady sisters, elbow some onto paths that lead out of the Solar System for good. Others are perturbed onto tight orbits where repeated blasts of the solar wind soon whittle their icy little skulls away to nothing. And some comets simply smack into the planets.

So the sky is not the only place to look for evidence

of comet showers. Surely there ought to be a scrape or a scratch on the surface of our own planet, even some thousands of them, to mark the last 700,000-year case of assault and battery. Certainly there are pockmarks, some no doubt put there by comets. But there are surprisingly few. Earth is not a good place to search for the scars of disfigurement. It repairs itself too quickly. Unlike the Moon, which carries the sorry tale of every punch in the lunar face, Earth toils away at her cosmetic labors ceaselessly, hiding her wounds beneath her veils of water and wood, until at last this "subduction," as geologists call the process of terrestrial renewal, pulls the mark right under, and nothing is left to the questing eye but the fresh and girlish skin of a blushing planet.

The Comet Spinners

So far, we have placed the comets out in their Oort cloud as the by-product of the great bang that helped to hammer the Solar System out of the mess and fury churning about the nascent Sun. This theory is not alone. A recent contender suggests that as the Sun drags the whole lot of us—planets, moons, and general debris—along through the chancy precincts of interstellar space, we pick up comets, or the makings of comets, from the clouds of gas and dust that drift around between the stars.

Or perhaps the comets were formed at the same time as the Solar System, but formed separately from the Sun. According to this model, proposed by A. G. W. Cameron of the Harvard-Smithsonian Center for Astrophysics, the proto-Sun evolved from the boiling chemistry of one interstellar cloud of material, the comets from separate and much smaller clouds of material. In other words, the comets were built right on the site they now occupy, not constructed at the center and delivered to where they are on the toe of some great solar boot. Cameron himself has now abandoned this theory, although not because he has decided to give up cosmology. He is still here, but with another theory:

Originally, comets orbited the Sun at a distance of, say, 300 AU, formed from the general stir of material available at the fringe of the accreted agglomerate that, still formless, surrounded the youthful Sun. The Sun had not yet begun its nuclear fusion and was just entering the T Tauri phase of its development. In the T Tauri phase, the stellar wind blowing off the young star tears through the prenatal Solar System like a hurricane, ripping loose whatever has not managed to attach itself to something massive enough to withstand the howling embrace. In only a few thousand years—mere seconds on the clock that times the universe—the T Tauri wind has blown fully half the original mass of orbiting stuff clean out of the front yard and beyond, out past the border at 200,000 AU and into interstellar space.

Suddenly (in universal terms) the protocomets find themselves orbiting a body with only half the mass it once possessed. The consequent drop in gravitation strings the comet out on a longer leash, that is, on an extended and more eccentric ellipse. Inching along at the end of its leash out at aphelion (the farthest point from the Sun), the comet is swiped off its lead by some passing star and joins the Oort cloud.

This is a lot to swallow, and although Cameron suggests another mechanism to account for the expulsion from the early Solar System of a mass equal to that of the young Sun, most astrophysicists prefer other models. The most accepted establishes the cometary assembly line in the vicinity of Uranus and Neptune, which would launch completed comets out to the Oort cloud. Here, too, there are critics. Why wouldn't Uranus and Neptune simply grab any local junk for themselves? There is no reason to believe that these two ladies are any more fastidious than their sister planets about what they eat. Another problem remains, however, even if we ignore Uranus and Neptune. All of these comets seem to end up in the Oort cloud. Where do comets come from to supply the inner cloud of Jack Hills?

The Twice-Collapsing Sun

Picture the proto-Sun, a vast, circular sheet of revolving gas and dust a trillion miles across which will become our Sun. In Hills's model, the comets are formed at the very edges of this sun, some 5,000 AU from the center. Here and there in the milling immensity of the disk, pockets of gas condense into brief and tiny stars, a million tons of mass stabbing a pinhead of light into the presolar gloom. Finally, the gravity of the mass at the center of the proto-Sun becomes so great that the star collapses. But not everything collapses at once. The shuddering slap of infalling material is only the first stage in the creation of the new star. The second stage occurs later, and over a longer period of time, as the luminous core pulls in the remaining material in a steady, majestic accretion that builds the star we see today.

Comets are almost certainly remnants of the very earliest moments of the Solar System. And yet we still do not know for certain just how these invaluable clues to our origins were made. But we are looking hard for answers. Enter the spy plane.

Mission over Russia

When most people think of the U-2, they do not think of astronomy. They think of a young pilot named Francis Gary Powers, a Moscow trial showcased to embarrass the United States, and a shadowy exchange at the border between East and West Germany that gained Powers his freedom at the price of a Russian spy.

The U-2 would simply be a gawky, spindly, fragile-looking aircraft were it not for its mission, and the tendency of its masters to paint it a sinister matte-black. The plane is little more than a rocket with a cockpit lashed aboard, lots of cameras, and a wingspan twice the length of the fuselage. It is designed for only one purpose: to fly where it is not supposed to fly, but so high that there is nothing those below can do about it except grind their teeth. It was the U-2 that spotted the missile installations in Cuba, sparking the 1962 confrontation between John Kennedy and Nikita Khrushchev that led to the withdrawal of the missiles. This was a great victory for the young president, and showed the value of the specialized aircraft. But earlier, it was the U-2 that presented the United States with an awkward international situation, the one that made Powers famous.

On the morning of May 1, 1960, at 5:20, Powers sat in the cockpit of his U-2 on the airstrip of the Turkish base of Incirlik, at Adana in southern Turkey. Incirlik was far enough from the Soviet border to prevent radar surveillance of what went on at the site. Even at that early hour, the tarmac baked under the furnace of the Turkish sun, and Powers's personal equipment sergeant stood atop his ladder holding a shirt over the pilot's head to grant at least some shade as Powers ran through his preflight checks. It was not until 6:20

A.M. that presidential approval for the flight clicked through the coded channels and Powers got the green light.

As soon as he was airborne, all contact with the base was over for Powers. His controllers would not take the slightest risk that the Russians, with their careful monitoring on their side of the border, might be alerted. The last sound that Powers heard from his own people, the signal that they acknowledged his mission as under way, was the double click of a transmitting key in the American command at Incirlik as radio contact was opened and then hurriedly closed. Not one human word. Just the two clicks. And then he was alone and heading for the Russian border.

By the time Powers broke cloud cover, he found himself to the south of the Aral Sea. Soon he knew that the Russians had picked him up. Far below, and traveling toward him on a converging course, was the vapor trail of a supersonic fighter. Soon after it had passed, another trail—Powers suspected from the same plane—overtook and passed him on a parallel course. Both times the trail was so far beneath his own altitude that Powers was untroubled.

It was four hours into his flight, over the city of Sverdlovsk, that Soviet technology caught up with American espionage. Sverdlovsk is the reincarnation of the little village of Ekaterinburg, where revolutionaries murdered Czar Nicholas II and his whole family in 1918. The Soviets changed more than the name when they redubbed the place Sverdlovsk. No longer a backwater village, it is now a center of considerable industrial and military importance to the Soviet Union. This makes it interesting to the United States, and Powers snapped away with his sophisticated cameras. He had just made a turn ninety degrees to the left and lined the U-2 up for another pass when his career as a spy pilot ended in a flash of orange light.

Today U-2s still fly. But not over Russia.

Comet Hawks and Vacuum Monsters

It is Moffett Field, thirty-five miles south of San Francisco, and the trim, familiar shape at the end of the runway glistens in the clear California sunlight. No longer clad in ugly black, this ER-2, an advanced version of its U-2 predecessor, is handsomely turned out in NASA's blue and white color scheme. It still looks a little perilous, as if the tips of its eighty-foot wingspan will wobble and scrape the ground when it begins to taxi. This impression does not last long. There is a booming roar from the small craft's powerful engine and in eight seconds flat the ER-2 is in the air and heading for its cruising altitude of 65,000 feet. So quickly does it climb that by the time it passes over the end of the runway, it is already two miles up.

The new version of the U-2 is ideal for its new mission: collecting comet dust. Astronomers hope to learn more about the origins of the Solar System by studying the microscopic grains of dust that drift to Earth through the atmosphere. The ER-2 collects these grains by deploying glass plates coated with sticky oil and flying through whatever is up there. The high altitude means that scientists have to wade through a lot fewer pollutants when they search the glass plates for grains from space. But even at those altitudes there are still terrestrial particles to sort out, tiny droplets of sulfuric acid from volcanoes, and titanium from airplane paint. Astronomers did not always have such elegant vehicles for collecting their dust. Before the graceful ER-2s were offered by the Air Force, dust collectors relied on an unwieldy device called the

Vacuum Monster. The Vacuum Monster was well named. He was basically a pump slung from a giant balloon. But he was a greedy fellow. Powered by 150 pounds of hydrazine rocket fuel, his pump gulped air at half the speed of sound. The air was drawn over a mesh of glass rods that collected the dust. Unfortunately, Vacuum Monster's parachute failed one day, and the 135,000-foot plunge spread him even more thinly across the ground than Humpty-Dumpty.

Besides the advantage of safer landings, the ER-2 can stay up for longer than a balloon-borne Vacuum Monster. After all, Clarence (Kelly) Johnson, the designer at Lockheed's supersecret "Skunk Works" in Burbank, California, had planned his original U-2 so that it could fly from one end of the Soviet Union to the other on one tank of gas.

Awaiting Halley's

The ER-2s will be sifting through the upper sky looking for bits of Halley's when it spreads its tail across our sky on this visit. But they will be only part of the welcoming fleet. There are plans for a regular regatta up there. The European Space Agency has plans to launch the probe Giotto to examine Halley's in 1985–86. The probe is named for the great painter who placed the 1301 Halley's comet in his masterful *Adoration of the Magi*. The comet represents the Star of Bethlehem. Giotto painted the *Adoration* soon after seeing Halley's himself. The probe Giotto will weigh in at three-quarters of a ton and is designed to plunge straight for the heart of the comet after spending some time watching its ultraviolet emissions and recording whatever else the Europeans can find to record. Giotto is expected to enter Halley's tail at 153,000 miles an hour, shielded against the blast of particles by double-skinned armor. A camera aboard the probe will transmit pictures as long as Giotto remains intact.

The Japanese Institute of Space and Astronautical Science (ISAS) plans to fire a probe named Planet A up to the comet. Planet A will fly past the comet about five days before Giotto reaches it. The Japanese will test their preparedness for the mission by boosting a test vehicle, MS-T5, into space six months ahead of Planet A. MS-T5 will test the ISAS deep-space communications network, but it will never approach closer than 10 million miles to Halley's. The intercept is solely the work of Planet A.

Planet A will be boosted directly out to Halley's orbit from the Kagoshima Space Center in southern Japan, and will approach to within perhaps 10,000 miles of the comet's nucleus. Planet A is a cylinder about three feet wide. Like the Europeans' Giotto, Planet A will spin-stabilize herself, which is to say she will spin like a top on location. A no-spin dish antenna will always aim at Earth.

The Soviet mission is a combined Venus and Halley's mission, involving two spacecraft, Vega 1 and Vega 2. The two identical spacecraft were launched from the Soviet Union in late December 1984; after swinging past Venus to drop landing probes, they will fly on to meet the comet, on March 8, 1986, for Vega 1, a week later for her sibling. The Vega craft each carries a platform loaded with cameras and experiments.

Where is America in all this frolic? Good question.

The first American plans were ambitious, and if carried out would have eclipsed the other action in a blaze of spectacular technology. A group of design teams coordinated by the Jet Propulsion Laboratory in Pasadena, California, produced a Halley's strategy that called for the launching of a spacecraft in 1981–82. Once in space, this craft would have either hoisted an enormous set of solar sails, harnessing the power of the solar wind, or else deployed a grid of vast solar panels constructed to collect enough energy to power a special ion-drive engine. This splendid and majestic contrivance was expected to sail its course for three, even four, years, gaining enough speed finally to allow it to match the velocity of Halley's comet when they met at a predetermined rendezvous in space. Flying in tandem with the blazing comet for several months, the ship would at last have closed to within a few miles of the surface, perhaps even landed on it. It would have been a glorious feat.

Pity the poor ship. She will not be built, at least not in time to meet the famous comet. No budget, say the bureaucrats; no voyage for the beautiful lady.

A Lady Named ISEE

Still, the United States is determined to keep an appointment with a comet, and even as you read this the venerable ship that will rescue the fortunes of America's astronomers is groping her way sightlessly through space, millions of miles away, heading for a rendezvous for which she was never designed. She is ISEE-3, NASA's International Sun-Earth Explorer-3, and she is a veteran lady of space. ISEE weighs 1,054 pounds, and her drum-shaped body was boosted into space on August 12, 1978.

ISEE was designed for a lifetime of only about three years, which of course she has already surpassed. Her mission was to orbit L-1, a point in space between Earth and the Sun where the opposing gravities are supposed to more or less cancel each other out. ISEE hauled a staggering burden of instruments up with her. Some of the antennae and booms project from her sturdy shape for 300 feet. Out at L-1, 900,000 miles away, ISEE went patiently about her tasks, firing off bursts of telemetry back to her masters on Earth, telling them about the solar wind, and giving them advance notice of shock waves heading for Earth. (Note: telemetry is the science of transmitting data from remote sources.) And when NASA's plans for a full-scale Halley's mission bit the dust of fiscal restraint, it was to ISEE that some of the shrewder astronomers turned their speculative eyes.

The first question was: Could ISEE do the Halley's job? Assuming they could get her off L-1 at all and on her way to meet the distant comet, would her radio be strong enough to reach from the comet back to NASA's Deep Space Network of huge dish antennae? The Jet Propulsion Lab ran the question through its computers. The answer: No. At first, gloom. And then someone remembered Giacobini-Zinner.

Giacobini-Zinner is not one of the magnificent comets, not a Halley's by a long shot. It is only about one-hundredth the size of Halley's and much less bright. But it is a regular visitor to the Sun, on an orbit of six years, and what's more, if ISEE could reach Giacobini-Zinner the United States could still scoop the world on a comet rendezvous, even if not with Halley's.

But the bureaucrats who dumped the whole Halley's mission into the garbage can were not any keener on Giacobini-Zinner. There were months of tense lobbying as scientists pressed for the mission. Their argument: It would cost only $5 million—peanuts in space budgets—and it would be an enormously prestigious first. While committees studied the proposal to go for Giacobini-Zinner, Dr. Robert Farquhar of the Goddard Space Flight Center in Greenbelt, Maryland, the man behind the Giacobini-Zinner proposal, ordered tests to see whether it was possible to find a trajectory to get ISEE out to the comet. The answer was affirmative. Finally the National Rescarch Council reported:

The exploration of a cometary environment for the first time is a scientific goal of the highest priority. . . . We recommend:

1. ISEE-3 should continue on its present trajectory to explore the magnetotail.

2. Subsequently, ISEE-3 should be sent to explore the tail region of the comet Giacobini-Zinner.

3. NASA should take steps to replace the monitoring function of ISEE-3 in a halo orbit.

What began next was a relay of the most complex set of instructions any spacecraft has ever been asked to execute. Late in 1983, Farquhar and his team fired ISEE's hydrazine-fueled rockets, sending her shooting away from L-1 to a new position in Earth's magnetotail, the field of magnetism trailed behind as Earth swings along her orbit. Then came the most critical move. On December 22, 1983, the Goddard Space Flight Center fired ISEE's jets again, plunging her to within a hairsbreadth of the Moon, no more than sixty miles above the lunar surface. This is the maneuver that gave ISEE the boost she needed, whipping her around the Moon like a rock in a sling and loosing her off into deep space on the trail of Giacobini-Zinner.

Giacobini-Zinner may not be a Halley's, but it is still an impressive object. Its coma is 32,000 miles wide and its tail plumes behind for 210,000 miles. As the brave little half-ton veteran plows her way into the sandblasting of the cometary bombardment, she will be pouring out whatever she finds in a torrent of telemetry. Back on earth American scientists and their guests from around the world will be watching ISEE's reports of her encounter with Giacobini-Zinner. The information should be invaluable to those countries still maneuvering their Halley's probes for the final plunge.

Once she was well on her way toward the rendezvous, ISEE was renamed, and she now voyages through space under the proud title of International Cometary Explorer, or ICE. The International in ISEE's new name is appropriate, signifying the remarkable amity that has grown up among the nations dedicated to intercepting the comet. ICE will be the first probe to brave the effects of a comet's tail, and when she barges into the streaming dust behind Giacobini-Zinner in September 1985, American space scientists will be relaying the stream of new data straight on to their colleagues in other countries.

The information will be awaited with some anxiety, for the probes aimed for the closest approach to Halley's—the Russian and European missions—carry highly sophisticated technology into the hazardous cometary environment. The Russian Vegas, four tons apiece and protected by multilayered foil and metal dust shields, will dive into the coma at velocities that are something like fifty times the speed of a bullet. Vega 1 will fly past Halley's at a distance of 6,000 miles from the nucleus. This is risky. While Vegas are built to withstand the impact of hurtling dust, anything the size of, say, a pea would be lethal, tearing into the spacecraft like a slug fired at point-blank range. But if Vega 1 does survive this brush, then Vega 2 will be directed to within 2,000 miles.

The European vehicle Giotto will attempt to approach to within 300 miles of the nucleus, filming with a 1,000-millimeter lens.

Other scrutiny of the comet will come from the American orbiter Pioneer, now on station orbiting Venus. Pioneer's position will enable scientists to watch Halley's even when it is out of sight from Earth behind the Sun. Another American contribution is the participation of the space shuttle. The mission Astro 1 is planned to monitor the activities of March 1986, when the probes converge on the outbound comet at its period of greatest display. Special ultraviolet scanning equipment and wide-field cameras will help the shuttle produce what will likely be the sharpest and most dramatic depictions of a speeding comet ever seen by man.

But the first word on comets will come back from ISEE/ICE, and the whole international community of space scientists was watching her in the fall of 1985 as she began to beat her way in toward Giacobini-Zinner, gaining for the rest of us a foretaste of Halley's.

CHAPTER 3

Murderous Space

There are killers out there, and this is the story of one of them.

The day the Apollo howled out of the sky, not one watcher on the face of this planet had detected its coming. The massive steam locomotive labored hard to pull the long train of freight and passenger cars through the endless, unpeopled forests of Siberia. It was the last day of June 1908, and the engine's cab seemed almost as hot as the furnace into which the fireman shoveled the black coal. The engineer ran an eye over the row of greasy gauges, cranking a handle over twice when he saw the pressure mounting in one of the pipes. A thin jet of steam sang its high song as the engineer leaned from the cab and watched it hiss through the valve he had opened.

On either side of the train the blank forest sat in the undisturbed ranks of centuries. Far behind the cab, in one of the dining cars, a few passengers still lingered over the remains of a heavy lunch. Talk was desultory, the bored, inattentive chatter of men with too much time on their hands. Russia was still the Russia of the Romanovs then. As the Trans-Siberian Express toiled its way across the vast continent ruled by the holy Czar, the world must have seemed as unchanging as the imperial state itself.

High in the upper levels of the atmosphere, the Apollo began to smoke, tearing itself to pieces as it plunged toward Earth.

Just behind the dining car, passengers in the plush coaches were reclining in their heavily upholstered seats. Some were snoring after their meal, their heads lolling against the linen covers fastened across the tops of the backrests. A few others read, and one or two stared out at the forest as it moved slowly past beyond the windows.

And then the Apollo hurtled itself into the lower sky.

The first person to see it, a young girl, said nothing. Her mouth open, she pressed back from the window into her mother, bumping hard. The woman turned to see what was the matter; then she screamed. By then, others had spotted the brilliant streamer, and shouts rang out along the train. What they saw was a brilliant blue fireball streaking across the sky at a fantastic speed. It disappeared over the tops of the trees, and a few seconds later the Apollo slammed into Earth.

It was only then, at the moment of impact, that the engineer ahead in the cab noticed anything. What he noticed was a series of loud bangs, and thinking there were explosions somewhere along the length of the train he throttled back and brought it to a slow halt. That is the impression the impact made upon a man working in a continuous blast of noise and heat. But the Apollo did not slam into Earth a mile or two from the tracks. *Its point of impact was 375 miles away.*

An Apollo is an asteroid or a spent comet whose orbit crosses the orbital path of Earth. They are wicked

75

little chunks of rock or metal, half a mile to a mile in diameter, and not a speck of light to mark their paths through space. They're bad news, all right, and that is why there are astronomers working right now at tracking and charting every one they can find. Left to themselves, *one in four of the Apollos whirling across our path right now will, sooner or later, pile into Earth at twenty miles per second.* There have already been close calls. Luckily for Earth, the only recorded impact in modern times, the Apollo that smacked into the valley of the Podkamennaya Tunguska River in Siberia that hot June day in 1908, was a baby. Nevertheless, *it exploded with the force of a hydrogen bomb.* If it had been of normal size, Earth would simply have been devastated.

So why is the Tunguska Event, as scientists call it, not better recorded? Why are there not more detailed and exhaustive records of the single occurrence that would have brought man to the very brink of extinction, but for a fluke of size? There would have been more details if it had happened in America. Or in Europe. Or even in the Soviet Union. But the little comet struck Earth in the middle of imperial Russia, and imperial Russia in 1908 was a nation virtually without a government.

Pale Emperor

This is how one American correspondent described the last of the Romanov Czars, Prince of Muscovy, successor to the iron Peter the Great. This is the man who ruled 135 million people.

The half moon of cobbled ground at the back of the Winter Palace in St. Petersburg is as fine an open space as any in Europe, and at its crest opens the great massive arch of the Marskaia, whence come the carriages of the great folk who drive from the Nevsky Prospekt. The arch is a majestic and ponderous span of terra cotta, surmounted with mighty men and horses in bronze, and from its cathedral shadows the carriage of the Czar emerged that day I saw him first.

It was a pale day of the moist, Russian summer, and the Czar had come from [the palace of] Peterhof to say farewell to a war-bound regiment. Through the arch and into the wan afternoon rode first a squad of cuirassiers of the bodyguard, a splash of cordial color flung suddenly into the scene. The meager light glinted on helmet and cuirass, and sparkled here and there on the drawn sabers and the accoutrement of the horses, and the party came down across the cobbles at a stirring trot, all sparkle and jingle, a fine and splendid thing to see. After them, another band of color, a chain, let us say, of strong hues—officers of the household, generals, admirals, and princes, and behind these, a small black victoria [carriage] drawn by two big stallions; and then more soldiers.

With a clatter of hoofs the party dashed down the way and into the great gates of the palace, but as they passed [one could] see a man who sat in the carriage, the man for whom all the splendor of arms and panoply was called into being. It was but the briefest glance, a mere peg on which to hang a first impression, but it told on me with an effect of dismay. Framed and overshadowed in the black hood of his carriage, I saw, bolt upright and motionless, a little figure immaculate and neat, with a face of dead pallor. Fair hair and beard duly dressed to a point failed to withdraw from it a quality of dollishness; an utter vacancy, the emptiness of soulweariness and futility, governed it altogether. Against its dark background, it stood forth as blank and white as paper, a thing awful in its corpse-like impassivity, yet pitiable, sorrow-stirring, and sad as a child in pain. The hands, I think, were crossed loosely on the knees, and I know that the eyes stared unwinkingly in front. It was a tragic effigy of weariness that the cuirassiers guarded, a body shrining a soul worn and distressed, a visible and warning token of the dread that stalks through Russia.

This was the man in charge. Prodded, pushed, and poked around as he was by whichever grand duke was most heartily ruining Russia at the moment, it is not surprising that Czar Nicholas was not as keenly interested as he might otherwise have been in the chunk of outer space that had landed out in the taiga, the Siberian coniferous forest. It was not until 1927, nearly ten years after Nicholas and his family died in a cellar in Ekaterinburg, that a serious expedition to the site was mounted.

Devastation in the Taiga

The following is the only account available of the expedition made, by L. Kulik. It is a translation from Russian made at the time by George P. Merrill of the United States National Museum,

The appearance at seven o'clock in the morning on June 30, 1908, of a "fiery body" of unusual brightness, rolling across the sky out of the northeast and falling down in the taiga between the Yenisei and Lena Rivers, north of the railroad line, was observed by a great number of people, mostly the native inhabitants, living in the basins of these rivers.

The fall of the meteorite [that is what Kulik believed the comet to be] was instantly followed by a column of fire rising skyward, by the formation of heavy black clouds, and by the most deafening, resounding noise far surpassing in its magnitude any thunderstorm or artillery cannonade. This was heard for hundreds of kilometers within a radius of the cities of Krasnoyarsk, Kansk, Yeniseisk, Nijneudinsk, and Kirensk on the Lena.

A terrific air-wave was formed that pushed ahead everything that it met in its way. The water in all rivers, lakes, and streams was raised up; people and animals were lifted by it and carried along.

The vibrations produced by the fall of the meteorite were detected and registered by the seismographs of the Physical Observatory at Irkutsk, where Mr. A. V. Vesnesenski, who was in charge of the observatory, calculated the epicenter of [what was thought to be] the earthquake to be located in the upper part of the Podkammenaya Tunguska.

The phenomenon produced considerable panic, especially among the natives living in the basins of the Yenisei and all the various Tunguska rivers, and adjacent part of the Lena River basin.

Several attempts made in 1908 to find the body of the meteorite were fruitless, as for some reason all parties were searching near the city of Kansk, and not in the locality determined by A. V. Vesnesenski, whose observations remained unfortunately unpublished. Gradually interest in the new meteorite died, and the whole matter was almost forgotten, except as a tale among the natives.

In 1927 Mr. L. Kulik attempted to find the exact location of the meteorite and led an expedition to the Tunguska region. Owing to the lack of funds and the extreme difficulties of transportation in the wilderness of taiga and tundra, the expedition was not altogether successful. However, Mr. Kulik was able to reach the area where the taiga bore distinct traces of the passage of the meteorite. An area struck by the meteorite is a water table between the upper part of the Podkammenaya Tunguska and its right tributary, the River Chuni. The area is largely covered with tundra in the process of formation, intersected by hills, small lakes, swamps, and typical tundra. The immediate area is surrounded by high naked hills, deforested by the falling meteorite. All the trees are still on the ground, their tops are spread out in fan-like fashion away from the central zone of the fall. Exceptions are noted only in the ravines or the gorges and deep perpendicular valleys, and also in a zone which can be considered as the "interference" zone. And even in these places the trees, in most cases, are scorched and though still standing, are all leafless and dead.

The zone where the heat effect of the meteorite is evident is considered by L. Kulik to be 30 kilometers in diameter [16 miles], and the area of the air-wave breaking the trees is 50 kilometers [32 miles] in diameter.

The central part of the "fire zone" is covered by shallow funnel-shaped craters, reaching in some instances many tens of meters in diameter and not greater than four to five meters [15 feet] in depth. The bottom of the craters is covered with swampy growth [twenty years after the impact].

Unfortunately, Kulik was unable to find the body of the meteorite or determine the depth to which it had sunk.

He believed that the meteorite of 1908 was an

aggregate [a swarm] of meteors, moving at a rate approaching 72 kilometers a minute [2,700 miles an hour]. Some of the aggregate undoubtedly exceeded 130 tons in weight. Hot gases (above 1,000 degrees Celsius) surrounded the meteorite and started fires before the meteorite had reached the ground and sunk into it, forming craters, uprooting the trees, and burning everything that could burn in the center of the fall.

Later, there were reports from a farmer named Semenov, whose home was fifty miles south of the impact point. Semenov said he saw a great light, or fire, in the north. When the light disappeared from the sky, the entire area was plunged into darkness and Semenov heard a deep, booming explosion which threw him down to the ground from where he was standing on his porch. When he regained consciousness, it was to find his house almost completely demolished.

Another man, a herdsman who kept a large herd of reindeer numbering about 1,500 in the area of the Podkammenaya Tunguska, returned to the area soon after the explosion. He had been away for three weeks, visiting in the region. Nothing was left of his reindeer except a few charred carcasses. His house was in ruin. The heat of the blast had melted even cooking utensils and iron tools.

On the evening preceding the comet's devastating visit to the taiga, observers in Sweden noticed a brilliant light appear in the sky. So strong was the light that long after sundown people could still read by it, and it remained in the sky until two o'clock the next morning.

That same night, *after* sunset, the sky over Britain began to lighten. The light increased until it was as bright as day. Meanwhile, seismographs on two continents were registering shock waves that pinpointed a massive "quake" in central Siberia. Following these events, the evening skies rewarded watchers with glorious sunsets in reds and yellows, blues and greens Similar sunsets were seen around the world after the eruption of Mount St. Helens; they are caused by the refraction of light through the vaporized material raised into the upper atmosphere by the cataclysm.

We now know that trees surrounding the impact o 1908 were instantly carbonized, that is, reduced to charred remnants in seconds. The column of fire above the principal point of impact was 5,000 feet across about a mile, and rose twelve miles into the Siberian sky. A circle almost twenty miles wide was set aflame the awesome cloud of smoke mushrooming above the burning land like the aftermath of an atomic blast. No living thing survived within eighteen miles of the impact. The crashing boom of sound rolled across Siberia for 600 miles in all directions. The devastation wrought by this relatively gentle tap from the boxing glove o. space is worth pondering. It is estimated that if the object had fallen to Earth five hours later, the city of Leningrad would now be a neat, circular bay off the Baltic. This is not too farfetched a notion to contemplate. There are plenty of scars on the face of Earth put there by a brutal cosmos. Some of these we *know* came from space; others are guesses.

Scarred Earth

Hudson Bay looks as if it might have been opened up by some celestial roundhouse punch; so does the Sea of Japan. The great circle in the Caribbean Sea described in part by the Lesser Antilles chain is suspiciously neat as is the Bay of Campeche off the Gulf of Mexico. What about a slug in the chops hard enough to leave the Pacific Ocean? It just might have been a similarly catastrophic event that wiped out the dinosaurs. But i needn't have taken anything as spectacular as that to terrify early man. Imagine even a minor impact, like the Tunguska Event, occurring in ancient Egypt. Perhaps it was exactly such an event that provoked the agony of distress evident in some of the papyri:

> Plague is throughout the land. . . . Blood is everywhere. . . . The land turns round as does a potter's wheel. . . . Oh! That the Earth would cease from noise, and tumult be no more! The land is entirely perished and nought remains. . . . The sun is veiled and shines not in the sight of men. . . . I show thee the land upside down. . . . None knoweth that midday is there . . . his [the Sun's] Shadow is not discerned. . . .

An early Tunguska?

What Was Tunguska?

There are plenty of outlandish theories to choose from to explain the Tunguska Event. One is that the great patch of coniferous forest was flattened by a flying saucer that landed, then took off again. No doubt the spacemen were disappointed to find they had plunked themselves down smack in the middle of Siberia. Unless they had journeyed a couple of hundred light-years in search of a few cords of firewood, one can understand why.

Another theory has the Siberian havoc wrought by a clump of antimatter, a sort of flying black hole. This is a tough one to sort out. Black holes are supposed to have appetites big enough to snap up whole planets. Or can a little black hole simply be filled in? Or—and this is another suggestion—might the black hole have shot into Earth on one side, Siberia, plowed through the center, and exited on the other side in the middle of an ocean, traveling so fast it didn't really have time to take a bite on the way through?

The most reasonable guess is that the visitor was a small comet, or even comet fragments. A shower of debris known to have been peeled off by the comet Encke about that time has become the favored culprit. The ignition of pieces of debris crumbling in through the atmosphere would account for the observed brightness in the sky in northern latitudes. These fragments— or, if they were not fragments of Encke, this small comet—became an Apollo when it fell into an orbit that crossed the orbit of Earth. This can happen to a comet that has made so many passes around the Sun it has burned off all there is to burn. Losing the "jet engine" effect of its volatile material, nudged by another Apollo, or perturbed by the powerful fields of Saturn and Jupiter, the comet declines into a solar orbit that intersects Earth's. Having no tails to mark their passage, these Apollos are hard to see. Until it is too late.

An Apollo can also be an asteroid, one of a belt of stony planetesimals orbiting the Sun between Mars and Jupiter. That is their normal habitat. But alas, they wander.

The Stealth of Apollos

In 1937 an Apollo named Hermes came barreling out of nowhere to whisper past our ears at the astonishingly close range of half a million miles. In Earth terms, this is like having a rock pegged at your head and coming close enough to raise hairs. If it had hit, the impact would have been equal to the detonation of 10,000 ten-megaton bombs. It would have blasted a hole in Earth fourteen miles wide, and when the dust had cleared away in two or three years there would have been no one around to write about it. And yet, although this deadly lump of iron and stone breezed past us at only twice the Earth–Moon distance, *not one astronomer on the planet saw it coming*.

Apollos are bad news.

Fortunately, most of the asteroids we know about lie in the belt between Mars and Jupiter, at a distance from the Sun of between 2.2 and 4 AUs. It is those asteroids that lie closer that are potential Apollos, bodies that approach to within 1.3 AUs of the Sun. Astronomers believe that any object that close will ultimately come closer, until finally it is drawn into an orbit that arcs across our own.

But Apollos are slippery devils, and are no sooner discovered than they usually creep off again into the anonymity of space, there to circle patiently until our backs are turned. The very first such asteroid to be discovered, the one originally called Apollo, was sighted, identified, and named in 1932 by Karl Reinmuth of the University of Heidelberg in Germany. It promptly vanished, and was not relocated again until 1973. Adonis, the second Apollo object to be identified, was discovered and tagged by astronomers in 1936. Like its predecessor, Adonis disappeared into space, and *remained hidden there for forty years, despite the fact that during that time it had crossed Earth's orbit thirty times*.

The term *asteroid* means "starlike," and that is how the larger members of the pack appear to astronomers. Although we have catalogued only about 2,500 asteroids, the asteroid belt contains approximately 400,000 objects ranging in size from pipsqueaks of a kilometer's diameter, to the giant Ceres with a diameter of 1,000 kilometers (600 miles). Many of these could become Apollos, perturbed by asteroidal collisions into Earth-orbit intersects. The vision of a monster like Ceres hurtling around the Sun on an intersecting path to

Earth's is the kind of thing astronomers may have been dreaming about when they wake in the middle of the night with sweat on their foreheads.

These days Apollos are hot stuff, and any astronomer picking out one of the speeding rocks on his photographic plates will rush to report his find to the International Astronomical Union Central Bureau for Astronomical Telegrams at the Harvard-Smithsonian Center for Astrophysics in Cambridge, Massachusetts. He had better rush; it will take him some time just to write out the address (see page 53).

Visions of Disaster

Exactly when an Apollo will strike Earth is strictly a matter of chance. But for eyewitness accounts of what it will do when it gets here, we have only to consult such authorities as George Griffith, whose *Olga Romanoff* supplied the imaginative climate for astronomical theorists at the turn of the century. Griffith's serialized novel describes a comet that comes to Earth to visit her just deserts upon Olga, who is tsarina of all the Russias and the wicked ruler of Earth. Olga has just tried to blow away the soldiery of the Republic of Aeria, and been booted from the field in disgrace. She is returning to her digs in Antarctica with her lover, Khalid the Magnificent. We are not told just what Khalid is magnificent at, though presumably it is not fighting.

The difference between the longitude of Aeria and Mount Terror had lengthened the last fateful day by nearly five hours, but now the end was very near at hand, and here, even on the very confines of the world, life had little more than four hours to live. To the north the whole sky was flaming out into indescribable splendors, and the long fire-streams radiating from the nucleus now seemed to be literally holding the planet in their clasp. Enormous meteors were bursting out from the heart of the flaming cloud and exploding without a sound in the ever-silent abysses of space.

She stood rooted to the spot by the weird and awful glories of the spectacle, and for the time being seemed to forget even Khalid and the indescribable dangers that were threatening them both. Instead of being daunted, her spirit rose as though in response to the splendors before her. She felt that she was standing upon Nature's funeral pyre watching the conflagration of the world she had ruined. Saving only Khalid there was not another human being within thousands of miles of her, and in her loneliness her soul seemed to expand and rise to a nobility that it had never known before.

She saw the utter insignificance and contemptibility of human strife which had been superseded and silenced by this majestic assault of the principal forces of Nature, and for the first time in her life she thought of herself and her sins with a disgust and shame that humbled her in her own eyes to the dust.

So she stood and watched, oblivious of everything but the celestial glories above and around her, until a frightening series of rapid explosions seemed to run roaring round the whole horizon. She looked up with shaded eyes toward the zenith. The central mass had suddenly become convulsed and expanded until it looked as if the whole sky had been transformed into an ocean of fire torn by storms.

Huge masses of many-colored flame were falling from it in all directions on the devoted [doomed] Earth, and as each of these entered the atmosphere it burst into myriads of fragments which fell in swarms until the blazing sky was literally raining fire over sea and land.

Fire-Cloud had at last invaded the outer confines of the Earth's atmosphere.

All this while there had been no change in the antarctic cold of the air, but soon after the first storm of explosions roared out, Olga felt a puff of warm tainted air blow past her face. Then came another and another, and then she heard what had never been heard before on the slopes of Mount Terror—the sound of running water. The snows were melting and soon there would come avalanche and deluge.

She hurried back into the council chamber, convinced that it was no longer safe to remain in the open air. She made the great bronze doors fast and covered them with layer upon layer of thick and heavy curtains. Every other opening into the chamber she closed up as tightly as possible. In the nature of the case they were compelled to trust to the supply of air already in it to last them through the ordeal.

Then she went and sat down on the divan by Khalid's side, and, taking his hand in hers, bent over him and kissed him on the lips saying—"Now we must wait for life or death together."

And so they waited—waited while the ages-old snow and ice melted from the bare black rocks under the fierce breath of the fire storm; while the ocean of flame seethed and roared and eddied about them, licking up the seas and melted snows and fighting with them as fire and water have fought since the world began; while the foundations of the Southern Pole quivered and rocked beneath their feet, and the walls of their refuge quaked and cracked with the throes of the writhing

EDMVND. HALLEIVS LL.D.
GEOM. PROF. SAVIL. & R.S. SECRET.

This is The Royal Society's own portrait of Edmund Halley, painted by Thomas Murray. The legend refers to Halley's attainments at the time, when he was the Society's harried secretary and holder of the Savilian chair of geometry at Oxford.

A segment of the Bayeux Tapestry, the Norman record of Duke William's victory at Hastings in 1066. The legend says "They wondered at the star," and someone is warning King Harold, soon to die, about it. The comet is Halley's comet of 1066.

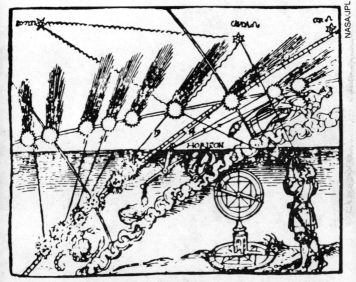

Representation of Halley's comet in A.D. 684, as published in the *Nuremburg Chronicles* in 1493.

The *Adoration of the Magi* by Giotto. The comet is Halley's comet of 1310.

Painting showing Halley's in 1682. Also pictured is a private observatory located on the city wall of Nuremburg.

An impact crater at New Quebec in the Canadian province of Quebec. This is an exceptionally young crater, only about 5 million years old, and a small one, no more than a couple of miles across. An Apollo object causing even a small crater like this one would have wrought terrible devastation in the surrounding area.

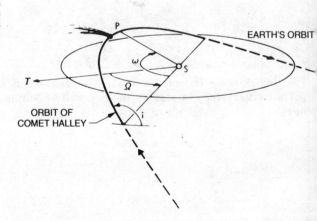

Halley's orbit for this apparition. The comet approaches Earth from beneath, if the star Polaris is considered "up."

ION TAIL

NASA/JPL

DUST TAIL

A comet actually has two tails. The action of radiant energy (sunlight) produces the dust tail, and the solar wind entrains the ion tail.

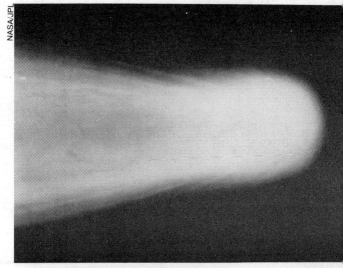

Head of Halley's comet, photographed May 8, 1910, from the 60-inch telescope on Mount Wilson.

Aerial view of the Very Large Array at Socorro, New Mexico. The 27 identical antennae here are 82 feet wide, can be moved around the site on standard-gauge railroad tracks, and are computer controlled to act as one massive antenna. The scrutiny of this massive facility will give us a better picture of Halley's comet than has ever been possible from a ground-based telescope.

Photo of Halley's from the Lowell Observatory, May 13, 1910. The bright spot is Venus, and the trailed lights of Flagstaff, Arizona, are visible.

The instrument that has revolutionized astronomy, the radio telescope. This instrument, with a diameter of 150 feet, is located in Algonquin Park in northern Ontario, and is typical of the telescopes with which astronomers probe space for trillions of miles.

The observer's cage in the mighty 200-inch reflector atop Mount Palomar. You can see the reflecting surface of the great mirror just in front of the cylinder that holds the observer. It was this telescope that first spotted Halley's comet on its latest return, picking it out of the night sky as far back as 1982.

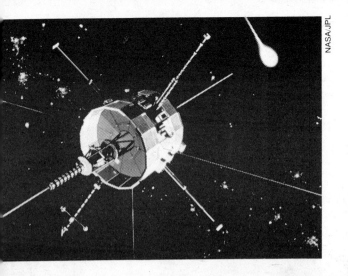

NASA's ISEE—International Sun-Earth Explorer—approaching the comet Giacobini-Zinner. ISEE has been re-named International Cometary Explorer—ICE—for her pioneering work.

Russian VEGA spacecraft. This is one of a pair of spacecraft that will make a combined Venus-Halley journey in early 1986.

The European Space Agency's Halley probe, Giotto. It is named for the great painter of the same name, who depicted Halley's comet as the star of Bethlehem.

A Frenchman and his wife believing they see the comet hurtling in to their destruction. Apparently they were not alone in their fear. The comet of 1857 excited a good deal of consternation among Parisians.

Les Planètes étant elles mêmes dans l'attente de la fameuse comète.

Not everyone fled in terror before comets. Honoré Daumier, the great French painter and caricaturist, found them hilarious. Here he depicts the planets Jupiter, Venus, and Mars enjoying a good laugh at the expense of Earthlings who feared the comet of 1857.

Earth, and cosmos was dissolved into chaos once more.

Nevertheless Olga, one of your hardier autocrats, survives all this cosmos dissolving. Khalid, alas, does not.

More Menace

No doubt Griffith's descriptions of cometary havoc were consumed with a good deal less merriment than we feel when we peruse them today. After all, a scant decade after *Olga Romanoff* was published, crowded churches and a brisk trade in comet pills testified to the apprehension developing over the visit of gentle Halley's. And as recently as 1857, the city of Paris was successfully stampeded into panic by an anonymous German astrologer who predicted that a comet would collide with Earth on June 13 of that year. In America, *Harper's Weekly* undertook a review of the prophecy, part of which detailed the history of the prediction:

Meanwhile a Belgian almanac-maker took the matter in hand, and predicted that this comet would strike the Earth on the 13th of June. From this gentleman's almanac-prediction have come all the rumors, the alarms, and excitements of the present year. A Paris correspondent writes: "For a fortnight we have not been able to step out without hearing the cry, 'Here is the end of the world! A full description of the comet of June 13, only one sou.' Eight women have miscarried; crops have been neglected; wills have been made; comet-proof suits of clothing have been invented; a cometary life insurance company (premiums payable in advance) has been created; and our 'Man About Town' has fancied himself walking the streets with a veritable blazing star—all because an almanac-maker of Liège thought proper to insert, under the week commencing June 13, 'About this time—expect a comet.'"

The combined attentions of the German astrologer and the industrious and cautionary Belgian were enough to rattle the populace of the French capital pretty thoroughly. But not universally. This is a report from the *Illustrated London News* three months before the comet was supposed to arrive:

At the last reception at the Tuileries, the coming comet was one of the subjects of conversation. Her Majesty the Empress seeing M. Leverrier, the well-known astronomer, among the guests, determined to make fun of the unsuspecting savant, and, feigning great alarm at the impending destruction of our globe, which this extravagant and erring luminary is, according to a German stargazer, to accomplish on the 13th of June, consulted him on the subject. M. Leverrier, to the great amusement of the guests, entered into a long refutation of this notion; and his embarrassment in endeavoring to avoid accusing Her Majesty, who ill-naturedly would not be persuaded, and his scientific enthusiasm, made the evening pass much more merrily than is usually the case with Imperial soirées—generally solemn stiff affairs.

Her Majesty the Empress was a regular barrel of monkeys, all right, no doubt supported in her amusement by the opinion of the scientist Jacques Babinet, who held that a comet belting into Earth would have no more effect than a fly smacking into the front of a railway locomotive. Tell it to the reindeer.

Looking for the Killers

Even today, so little is known of Apollos that even guesses about the number of them are just that—guesses. In 1963 Ernst Opik of the Armagh Observatory in Ireland made the first attempt at estimating the Apollo population. At the time astronomers had tagged only ten of them for sure. Opik's figure: at least forty-three, and possibly many more. Certainly Opik was not unrestrainedly lavish when it came to awarding population figures. But Apollos then were pretty exotic phenomena, even more persistently elusive than they are now. A much more recent guess is that the total number of Apollos with an absolute magnitude brighter than +18—in other words, Apollos with diameters of at least one kilometer—is somewhere between 450 and 1,050.

We know that a typical Apollo will intersect Earth's orbit about once every 5,000 years. So if we take the worst case of that population guess, 1,000, that means there is one of those very-hard-to-see little rocks arrowing across our path every five years. This would be nerve-racking were it not for the comforting fact that our orbit around the Sun is 950 million miles long, and laid-back Earth idles along it at a serene 67,000 miles an hour. There are so many places for us to be in our orbit when an Apollo is crossing it that the chances of any single Apollo's catching us where it wants us is something like once every 200 million years. There is, however, more than one Apollo out there hacking away at us in the leisurely time scheme of space. This reduces the odds to about four collisions every million years, or once every 250,000 years. So when are we due for the next blow?

Other than the Tunguska Event, which was a smaller-than-typical Apollo and counts only if you are a reindeer, the most recent mark from an asteroidlike body is Meteor Crater in Arizona. Scientists date Meteor Crater at 25,000 to 50,000 years old. But it is only a little over one kilometer, or five-eighths of a mile, wide, and an unimpressive 600 feet deep. Meteor Crater is the work of a lightweight. Keep looking!

The most recent crater large enough to have been excavated by an incoming Apollo is the depression holding Lake Bosumtwi in Ghana, West Africa. It is 1.3 million years old. Thus, we are overdue for an Apollo visit, and long overdue at that. Two, at least, have tried to pay that visit.

In the early morning of February 28, 1982, the astronomer Hans-Emil Schuster at the European Southern Observatory at La Silla, Chile, spotted an asteroid which he tentatively named 1982DV. Tracking it, he calculated an orbit that begins out in the asteroid belt between Mars and Jupiter and ends just barely short of Earth's orbit. Although 1982DV, a hefty three kilometers wide, was not technically an Apollo because its orbit did not quite reach ours, astronomers believe that eventually it will. This last time it came to within 14 million miles, a close brush by space standards.

On the very same morning that Schuster found 1982DV, Eleanor Helin at the Palomar Observatory in California spotted a smaller, one-kilometer-wide asteroid, this one an Apollo. Named 1982DB, it approached to within 10 million miles of Earth. Moreover, *this Apollo is in an orbit that will cross ours once every 1.9 years*. But even more disturbing than that, the little Apollo was not picked up until after it had already made its closest approach. What this indicates is that even with astronomers scouring the heavens with their advanced radio telescopes, we still might never see the bullet that kills us. Some people take this seriously. Here is what the Massachusetts Institute of Technology estimates would happen if one asteroid, a threatening fellow by the name of Icarus, collided with our planet.

Project Icarus

If Icarus were to smack into the Atlantic a thousand miles east of Bermuda, tidal waves moving at 400 to 500 miles an hour would simply wash away the Bahamas and all the resort islands of the Caribbean. Cuba would be reduced to a shattered mess. Florida would be swamped. The city of Boston, 1,500 miles away from the point of impact, would be battered by tsunamis 200 feet high.

If Icarus struck dry land, the crater would be fifteen miles across and three to five miles deep. A shock equivalent to half a trillion tons of TNT would lift a million tons of Earth's crust into the air as pollution. Our skies would be black for four years. Earthquakes and winds of incalculable magnitude would savage the planet. Our civilization would sink beneath a blanket of choking dust.

Farfetched? Not at all. Those data on the probable disastrous effects of an impact with Icarus emerged from an MIT project in systems engineering called Project Icarus. The purpose of the project was to save planet Earth from devastation by an impending strike of the Apollo Icarus. In all respects the project was executed as if the threat were real, as indeed someday it will be. This is part of the notice that appeared on the bulletin boards throughout MIT in the spring of 1967:

Clearly, Icarus must be stopped. No effort or funds will be spared in carrying out the detailed plan to be developed by the crack team of engineers and scientists assigned to the project. The problem solution may utilize a rocket to intercept the [Apollo] asteroid and nudge it from its course. Alternatively, it may be better to reduce it to rubble with a nuclear warhead. Multiple booster vehicles and rendezvous may be necessary to meet payload requirements. Gemini and Apollo [program] hardware may be necessary, and may be utilized if a manned space system enhances probability of success.

The students, at first inclined to treat the project with cynicism, soon settled down to the fervor that befitted the elite technocrats designing the Savior scenario for Earth. The plan they eventually evolved called for a series of Saturn V moon rockets to boost the payload into space. Launched in the last thirteen days before impact, the massive rockets would deliver six payloads to intercept the approaching Apollo. Each payload would consist of a hydrogen bomb with a destructive force of 100 megatons. The purpose of the deterrent was either to fragment the Apollo or to deflect it from its course entirely.

In a final evaluation of Project Icarus, Professor Richard W. Heldt of MIT's Department of Aeronautics and Astronautics wrote that a grateful Earth could:

expect an 86 per cent reduction in damage due to the efforts of the Project team. Perhaps even more encouraging is the 71 per cent chance of no damage at all because of the 71 per cent chance of deflection. But regardless of the probability associated with the success of Project Icarus, its cost and sociological impact are clearly insignificant in the light of the staggering alternative: disaster.

Filling the Ranks

The astro-ecologist need never worry that by blowing these fellow citizens of the Solar System out of the sky we will be exterminating the Apollo race for all time. New ones will always appear to take the places of their vaporized brothers.

Comets will replace the holes in the ranks.

Most of a comet's life is spent far out beyond the most distant planet. We see them only when a passing star shunts them into an orbit whose perihelion is fairly close to the Sun. In most cases, some great brute of a planet, like Jupiter, will elbow the comet off into space again before it has burned off much of its load of ice and snow. But sometimes the planets act in the opposite way, and pull comets into even tighter orbits around the Sun, perhaps hauling them into orbits with periods as short as seven years. For millions of years the comet orbits the Sun, logging perihelion passage after perihelion passage, its mass of volatile material slowly and steadily blowing away in the currents of the solar wind until all that is left is a black, invisible rock. One little sideswipe with any of the other pieces of stone whirling around the Sun, and our comet has become an Apollo. If this happened once every 65,000 years, it would create enough new Apollos to replace the ones that bash each other apart, splatter into Mars or Venus, and menace Earth.

Mind you, all this is pretty tame stuff compared to what a comet can really do when some capable Earthling starts examining the records. In 1950 Immanuel Velikovsky, an eccentric psychiatrist with a vivid imagination, set the scientific world a-splutter with the publication of his fanciful book *Worlds in Collision*.

According to Velikovsky, a great comet winged in our direction by the planet Jupiter was responsible for

Noah's flood and for the parting of the Red Sea that allowed Moses to escape the pursuing Egyptians. While blustering about near Earth, the comet also managed to dump a load of frogs on us. Fortunately for the frogs, the comet laid in a supply of flies. If you had been around when the comet was performing all these feats of natural history, you would have caught the action yourself, as Velikovsky insists our ancestors did.

Dr. Velikovsky is dead now, and so it is probably best to leave his memory in the anxious care of the astonishingly numerous band of devotees who swallow all this rubbish. Let us merely note that penury was not the cause of Dr. Velikovsky's death. That says something for his business acumen. About his scientific abilities, you may judge for yourself from the following:

At the same time that the seas were heaped up in immense tides, a pageant went on in the sky which presented itself to the horrified onlookers on earth as a gigantic battle. Because this battle was seen from almost all parts of the world, and because it impressed itself very strongly upon the imagination of the peoples, it can be reconstructed in some detail.

When the earth passed through the gases, dust, and meteorites of the tail of the comet, disturbed in rotation, it proceeded on a distorted orbit. Emerging from the darkness, the Eastern Hemisphere faced the head of the comet. This head only shortly before had passed close to the sun and was in a state of candescence. The night the great earthquake shook the globe was, according to rabbinical literature, as bright as the day of the summer solstice. Because of the proximity of the earth, the comet left its own orbit and for a while followed the orbit of the earth. The great ball of the comet retreated, then again approached the earth, shrouded in a dark column of gases which looked like a pillar of smoke during the day and of fire at night, and the earth once more passed through the atmosphere of the comet, this time at its neck. This stage was accompanied by violent and incessant electrical discharges between the atmosphere of the tail and the terrestrial atmosphere. There was an interval of about six days between these two close approaches. Emerging from the gases of the comet, the earth seems to have changed the direction of its rotation, and the pillar of smoke moved to the opposite horizon. The column looked like a gigantic moving serpent.

When the tidal waves rose to their highest point, and the seas were torn apart, a tremendous spark flew between the earth and the globe of the comet, which instantly pushed down the miles-high billows. Meanwhile, the tail of the comet and its head, having become entangled with each other by their close contact with the earth, exchanged violent discharges of electricity. It looked like a battle between the brilliant globe and the dark column of smoke. In the exchange of electrical potentials, the tail and the head were attracted one to the other and repelled one from the other. From the serpent-like tail extensions grew, and it lost the form of a column. It looked now like a furious animal with legs and with many heads. The discharges tore the column to pieces, a process that was accompanied by a rain of meteorites upon the earth. It appeared as though the monster were defeated by the brilliant globe and buried in the sea, or wherever the meteorites fell. The gases of the tail subsequently enveloped the earth.

The globe of the comet, which lost a large portion of its atmosphere as well as much of its electrical potential, withdrew from the earth but did not break away from its attraction. Apparently, after a six-week interval, the distance between the earth and the globe of the comet again diminished. This new approach of the globe could not be readily observed because the earth was shrouded in the clouds of dust left by the comet on its former approach as well as by dust ejected by the volcanoes. After renewed discharges, the comet and the earth parted.

Thank God for that!

CHAPTER 4

The Star Mapper: Edmund Halley

Who was the turbulent man who fastened his name forever to a lump of blazing snow? He drank brandy like a sailor and swore like a sea captain. After all, he *was* a sea captain. He was also an inventor, a navigator, an astronomer, a diver, and a life-insurance actuary. And he was a mathematician who dragged into being one of the greatest documents of science of all time, Newton's *Principia,* when Isaac Newton, that most retiring of men, would have happily left it blowing in dusty neglect among the mess of papers tumbled on his desk.

Edmund Halley was born to a prosperous father in London in 1656. There is some doubt about the exact date of his birth, since the universal keeping of baptismal records had fallen off along with the head of King Charles I, and Oliver Cromwell's Puritan Commonwealth was anticlerical. Happily for Halley, who was to receive much in the way of assistance and support from his sovereigns, the monarchy was restored when he was four years old.

Edmund Halley, Sr., was a soapmaker and salter, and could well afford to give his bright son the education he so clearly deserved. In 1673 Halley went to Queen's College, Oxford, a youth of barely seventeen years but already a keen astronomer and thoroughly grounded in the mathematics of his day.

How's That?

Nothing excites the pedant in man so much as the trifling detail, so let us clear one such obstacle from the pathway before we proceed further. How is Halley's name pronounced?

The most common pronunciation is Hay-lee, perhaps suggested by the former pop group Bill Haley and the Comets. Junk this one. It is almost certain that Halley did not pronounce his name that way. Then there is a case made for the pronunciation Haw-lee, on the grounds that Halley was sometimes addressed with that spelling and that he sometimes signed himself that way. But this is no proof, for Halley's was an age of more adventurous spelling than our own, and he signed himself with several different versions.

In England, most people who share the same name as the great astronomer pronounce it to rhyme with *Sally,* and that is the pronunciation which is used today by scientists. But, by all means, take your pick. It is unlikely that it would have bothered Halley much when he was alive, and it certainly won't now.

The Student's Universe

Halley did not arrive at Oxford packing a tin case full of amateur scrap hammered together in his basement. His father, always generous where his son was concerned, had supplied him well. Any scholar at the university would have envied Halley's telescope, which was twenty-four feet long. His sextant was a proud device two feet in diameter, and with this superior equipment the young man was soon making those painstaking observations that characterized his whole career. But no man begins any endeavor on his own, and Halley was blessed by being able to clamber onto the shoulders of the brilliant men who were liberating astronomy from the old ideas of a thousand years of history.

The prevailing vision of the universe in Halley's youth was still the one bound tightly in the pages of the *Almagest*, the work compiled by Ptolemy around A.D. 150. In the Ptolemaic scheme, Earth was the center of all things, with the Sun, Moon, and planets moving about it in perfect circles, and the stars fixed forever into spheres of crystal. But this idea was giving way, and in England many astronomers were already accepting the theories of Copernicus, who placed the Sun at the center of the observable sky.

Another man whose discoveries were exploding the orthodox tenets of the time was Tycho Brahe. Brahe lived on the Danish island of Hveen, in a castle named Uraniborg—the City of Uranus. Dressed in magnificent garments, Brahe ruled his own realm as if he were a god himself, treating his retainers like slaves, and plundering the treasury of his royal Danish patron.

Brahe developed observational techniques that were astonishingly accurate, and through these techniques determined that a large comet he saw was actually *beyond* the orbit of the Moon. This was startling stuff, for until then astronomers had assumed that comets occurred only in the lower sky—that is, the sky "beneath" the Moon. But if the comet was not a denizen of the lower, sublunary regions, and instead traveled far beyond the Moon, where did it come from? If it came from beyond the stars, as was now apparent, how did it get to where it was seen by men, for were not all the stars held firmly in place by the spheres of crystal? If there were really crystal spheres up there, then the comet must have broken through them on its way in. And if a comet had smashed through all that crystal, why were the stars still up there where they had been all along?

Good-bye, crystal spheres. They were heaped into the same trash can as the elaborate Earth-centered universe.

A disciple of Brahe's, and the man who ultimately inherited the volumes of measurements taken by the Danish astronomer over the decades, was Johannes Kepler. When he was not being plagued by the nobility to discharge some of his more irksome duties—like casting horoscopes for members of the royal court in Prague—Kepler tackled the problem of explaining the revolution of the planets around the Sun. Over ten years of dogged research Kepler found that circular orbits simply would not work, and eventually proved that planetary orbits were in the shape of an ellipse.

One of the most intractable objections to the Sun-centered universe was tied to the Moon. If Earth moved around the Sun, as was asserted, then why didn't the Moon fall behind and simply drop away into space as our planet sped off on its orbit? The answer was supplied by Galileo Galilei.

Galileo was then still experimenting with his copy of

the Dutch telescopes. His own copy of these new and advanced instruments had been made under his supervision by the expert glassmakers of Venice. With his brand-new telescope, Galileo discovered that Jupiter, too, had moons, and further, that these moons orbited the giant planet. They were, in fact, satellites. If so for Jupiter, why not for Earth? Problem solved. The Moon stayed right where it had always been, patiently circling Earth.

Another of the great men whose work Halley would have studied was René Descartes, the French philosopher. It was Descartes who, beginning with the premise "I think, therefore I am," erected one of the most subtle and durable philosophical structures of all time. Part of it was his conception of the vortex, a theory of the universe that divided all matter into particles, some heavy, others light.

It was also Descartes who developed the system of coordinate geometry still in use today, more than 300 years after his death. Using Cartesian geometry, mathematicians can plot the position of one object by reference to the known position of another.

All of this was of immense practical concern, for the great maritime nations of the seventeenth century were all hotly competing with one another for colonial possessions and influence, and any improvement in astronomy that would give one an edge in celestial navigation was eagerly sought. So it was not just because he loved learning that King Charles II established the Royal Observatory on Crown lands in 1675. And when the Reverend John Flamsteed took up his position as the first astronomer royal, setting up at Greenwich on the Thames below London, he knew that the solutions of navigational conundra—like the means of establishing longitude at sea—were matters close to the interests of the realm.

The Passionate Recorder

While other students in the quiet university town of Oxford were closeting themselves with their dons in the placid quadrangles of scholarship, Halley was nightly scanning the heavens with his telescope and pouring out his observations in a stream of letters to the new astronomer royal. This was no uncertain teenager, and Flamsteed, ten years older than Halley, appears to have recognized the youth's mettle. Many of Halley's measurements (nothing less than corrections of the tables then in use among scientists decades his senior) were published by Flamsteed in the *Journal of the Royal Society*. But Edmund Halley was not content to huddle over the charts, correcting a figure here and changing a decimal there. He craved a greater challenge, and he conceived a daunting one: he would chart the southern sky.

Far down in the South Atlantic, at latitude 15°55′ south, lies a tiny chunk of leftover volcano which is the British possession of St. Helena. In recent years the rugged little island served as a strategic staging point for the British assault to reclaim the Falkland Islands from Argentine invaders. In Halley's day St. Helena served British interests in another way. It was a principal transit station for ships of the powerful East India Company, both as they returned from India and as they sailed out for the Cape of Good Hope and the riches that lay beyond.

Once Halley had settled upon St. Helena as the location for his undertaking, he began to pull hard on the only string available, the one attached to John Flamsteed. Impressed with his young correspondent's project, Flamsteed went to Sir Joseph Williamson, the secretary of state, and at length His Majesty was persuaded to send a letter to the East India Company recommending Halley's plan.

Certainly, a recommendation from the king was a sharp enough nudge that it would be felt by even the most exalted of ribs. But probably the canny merchants who governed the East India Company appreciated, too, the navigational boost the young man's labors would produce.

Halley went handsomely endowed. His father, always indulgent, was nothing short of munificent. He provided his son with an allowance of 300 pounds a year. To appreciate the worth of this, it was six times as much as the great diarist Samuel Pepys was then making in an important position at the admiralty, and three times the salary of the astronomer royal.

South to St. Helena

Halley boarded the stout little vessel *Unity*, likely a product of the company's own shipyards at Deptford on the Thames. Off they went on their 5,600-mile journey down the Atlantic, bucketing out of the English Channel and into the broad ocean for a voyage that would last three months. Secured in the holds were instruments either specially designed to withstand the pounding of the ocean or stowed with exceptional care. Among them were the twenty-four-foot telescope and a new sextant, this one with a radius of five and a half feet, fastened to a steel frame.

The *Unity* reached St. Helena in February of 1677, and Halley and his traveling companion, a Mr. Clerke, set about lugging their gear up the 2,700-foot-high slope of Diana Peak, the highest elevation on the island. Unfortunately, the governor, Gregory Field, was as sour a civil servant as ever plagued a citizen, and was nothing but a hindrance to Halley. (For those who believe in such things, this unhappy relationship may seem to have been prophetic. Two centuries after Halley encamped on the peak, his outpost would overlook the final prison of Napoleon, Longwood Farm. And by some accounts the last days of Napoleon's own tenure on the island, which ended only with his death in 1821, were dogged by a governor every bit as hateful as Field, the infamous Hudson Lowe.)

Besides the governor, Halley had to battle the weather. Where he expected to find clear skies, he found clouds and rain and mist. Only by making use of every break in the oppressive cover could the young astronomer log his observations. But Halley was a bulldog of tenacity, and the red eyes of sleeplessness were not strong enough enemies to keep him from his task. When he returned to London, it was in triumph, and he published his great *Catalogue of Southern Stars*. It was the work of a master, and as a master he was hailed; Flamsteed called him the Southern Tycho. But there was something more than praise that Halley wanted, and if you conclude that he used a little cunning to gain it, well, no one ever called Edmund Halley a simpleton.

The problem was this: Halley had no degree. Then as now, the little string of letters a man could pin behind his name made a great deal of difference to the people in charge of the kinds of jobs Halley wanted. Halley, having abandoned Oxford for St. Helena before his degree course was finished, had failed to satisfy the residency requirements demanded of the university for any degree candidate. Hailed as a brilliant astronomer he might have been, but the guardians of the rule book were unimpressed. No degree for Master Halley.

Naturally, Halley began to reach once more for convenient strings to pull, and by now his success had helped attach them to a broader range of useful people. Halley began to tug away, but he didn't leave the matter there. When he was preparing his planisphere, or celestial chart, for presentation to the king, he sagely decided to invent a new constellation out of stars he simply borrowed from the constellation Argo, which already had enough of its own. This new constellation Halley named Robur Carolinum (literally, the Oak of Charles), commemorating the tree in which the king had hidden to escape capture by the forces of Oliver Cromwell after the Battle of Worcester in 1651.

The king wrote to the vice-chancellor. Halley got his M.A. The same year, the young star mapper was elected a fellow of the Royal Society. He was twenty-two years old.

Into Europe

Throughout his life, Halley was a man of the most robust eclecticism. He loved foreign travel. He loved foreign food. He loved foreign drink. And he loved foreigners. As well fitted to diplomacy as to science, he was a natural choice to heal a quarrel that was dividing the community of scholars.

Johannes Hevelius was one of the titans of observational astronomy of his day. From his observatory at Danzig, where the Vistula flows into the Baltic Sea, he charted the stars through open sights, peering through rudimentary cross hairs along a sighting arm. This was the nub of the dispute. Some of his fellows in England (Hevelius was a member of the Royal Society, although a German) questioned his measurements, claiming that the open sight was anachronistic and less accurate than the telescope. Hevelius himself had a telescope, a monster that was 150 feet long. But he used it only for viewing the planets and the Moon, maintaining the superiority of the open sight for stellar charting. If this sounds like a gentlemanly difference of opinion, it was not. The debate was bitter, and charged with insults and acrimony no less biting for being phrased in Latin.

Happily, the young Englishman and Hevelius hit it off well. Indeed, they had a lot in common, for both of them were the sons of rich fathers. More than that, Hevelius's father had been a brewer, and the astronomer still ran the family firm. It may be that this stroke of fortune helped endear the older man to the convivial Halley. When Halley left Danzig, it was with warm praise for the German's work, and it seemed for a time as if the breach between Hevelius and his British colleagues had been healed.

This changed. Soon after Halley left Danzig, a fire caused by a careless servant devastated the famous observatory while Hevelius was out of town, wrecking his instruments and burning whatever precious books were not stolen. Hevelius, shocked and suspicious, began to suspect that Halley had come to Danzig to spy upon him. Full of wrath and believing himself a victim, the German began to publish it abroad that Halley had made his voyage to St. Helena at the express urging of Hevelius.

This angered Halley, who replied that he had supported the German only out of a wish to console a "peevish" old man. The debate over telescopes versus open sights did not end until Hevelius's death in 1687. It was telescopes from then on.

Halley went abroad again in 1681, this time capping his education with the grand tour. Traveling was a grueling business in those days, by horseback and cart and punishing coach. When Halley and his friend Robert Nelson, the son of a merchant in the Orient trade, rolled into Paris, it was not to turn their luggage over to some porter in the Gare St-Lazare. Travel was the pursuit of a rugged soul housed in a sturdy body.

Although the appearance of a bright comet sent

Halley running to the observatory at almost the moment he set foot in the French capital, he recorded a breadth of activity that shows what a man of parts he was. Starting at the observatory, he set off to pace the length and width of the city, concluding that although it contained fewer houses than did London, it must be more densely populated, for the records established that there were more people being christened and more being buried. This was not the only conclusion he drew:

> . . . and the Christnings farr exceed ours, having been almost 19000, when we have ordinarily 12 or 13000, here they likewise take an account of the weddings which were 4470 last yeare or a quarter part of the Christnings very neare now in these weddings halfe as many were married as were borne; and not more; it will from hence follow, supposing it alwaies the same, that one halfe of mankind dies unmarried, and that it is necessary for each married couple to have four Children one with another to keep mankind at stand. This Notion Occurred whilest I was writeing. . . .

This kind of speculation, and the conclusions drawn, are not much different from the more lucrative toil of actuaries hunched over desks in air-conditioned offices, poring over the latest figures from the Bureau of the Census and calculating premiums!

Marriage and Tragedy

Edmund Halley was back in London by January of 1682, in time for his father's second marriage. Within three months he was married himself, to Mary Tooke, the daughter of an official of the Exchequer. Whatever else it may have done for him, marriage did not dissipate Halley's energies, which thundered on. It was now that he proposed the first of those audacious theories that led one twentieth-century geophysicist to call him "a leading member of that single generation which . . . discovered the content and invented the technique of physical science."

Briefly, Halley decided to hang on what today would be considered the skimpiest web of data the proposition that Earth is composed of both an exterior sheet and a core, each of which has its own set of magnetic poles. Further, he said that the shell rotates faster than the core, by one-half of a degree a year. And by God, he was right. This is amazingly close to what Earth scientists believe today.

Edmund Halley, Sr., was murdered in March of 1684, his body found by a river near Rochester, in Kent, five days after his wife reported him missing. He was so badly disfigured by his murderers that it was only his shoes—all that was left of his clothing—that enabled the family to identify him. The death must have hurt Halley, who was close to his father, as generous and unflagging a supporter of genius as any brilliant son ever had.

Dragooning Newton

Edmund Halley was twenty-seven years old, a respected fellow of the Royal Society with a record of practical and theoretical achievement that many of his elders could only mutter at in dismay. But Halley was just starting. And he was beginning to probe an area of astronomy that would lead him, and swiftly, to one of the greatest and most selfless acts ever recorded in the thorny chronicles of man: the single-handed and relentless badgering of Isaac Newton until that superstar of his age finally coughed up his staggering achievement, the *Principia*.

By 1684 Halley was bending his restless brain to the study of Kepler's laws of planetary motion. Sweating away at his laborious calculations, he discovered that the Sun's power of attraction on a planet decreased with the distance of the planet from the Sun in a ratio tied to the square of that distance. If you doubled the distance of a planet from the Sun, the Sun's power over that planet decreased by four times. If you tripled the distance, it decreased by nine times. And so on. But Halley could find no way to *prove* this scientifically.

In August of 1684, he stuffed his voluminous calculations into a sack and headed off to Cambridge, where the retiring Isaac Newton dwelt in the seclusion of Trinity College. Seldom has a more reluctant figure been dragged so unwillingly onto the stage of universal fame, and yet any number of Newton's achievements would have guaranteed him a secure place in history. This shy and reticent man, a farmer's son, had already invented the reflecting telescope, an improvement on the conventional refractors. He had formulated the theory, far in advance of his contemporaries, that white light was actually made up of the presence of all colors of light.

When Halley presented his findings, Newton agreed with them. Moreover, he claimed that he had already worked out the proof that had eluded Halley. But there was a catch: he couldn't find it. So casual was Newton about his work that he had simply lost it among the papers tangled in piles about his study. Unworried, he promised Halley that he would just sit down and work it all out again. He did, and sent it off to Halley in London.

Examining Newton's work, Halley realized that the mathematician had evolved nothing less than a whole

111

new scheme of thought. Having not only the wit to detect it, Halley also fortunately had the resolve to see that it was freed from the hands of its timid creator and revealed to the world. Manfully tackling the project, he persuaded Newton to agree to put all of his work together into one monumental opus. However, there was one snag. In order to work out his laws, Newton had invented, single-handedly, differential and integral calculus. Halley realized that if Newton were to use his new methods to expound his theories, no one would understand, and the great laws would be greeted by a sea of blank faces. Newton therefore agreed to produce the work using conventional geometry to support his contentions. It took him eighteen months of hard slogging, and Halley hovered over him every page of the way.

The Wigless Clerk

While all this was going on, Halley took a new job, as clerk of the Royal Society. Just why he took the position is a bit of a puzzler, since the pay was low and he had to resign his prestigious fellowship in order to accept a salaried position. Even worse, he was henceforth required at meetings of the society to sit "uncovered [without his wig] at the lower end of the table." But exactly where he sat, and whether or not he did so draped in a wig, do not appear to have been the kind of trivia that much bothered Halley.

Everything interested him. If he was not discussing the effects of "Oyle of Turpentine" in breaking up clay deposits, he was corresponding about medical matters:

The child you mentioned to have seen with six fingers on a hand and as many toes on each foot, is a great curiosity, especially if they be so contrived that the hand be not thereby made less fit to do its office. Nor is the quantity of Water found in the Dropsicall maid less prodigious, it being hardly conceaveable how the Muscles of the Abdomen should be distended to so great a Capacity.

You may be forgiven for wondering whether the "Dropsicall maid's" belly in fact was full of water, or whether it contained some other matter which she wished to conceal—like a baby.

The Royal Society was meanwhile encountering one of those problems that seem to plague every organization but the Rotary Club: they were short of money. So broke were they that Halley had to take one whole year's salary in copies of a book called *The History of Fishes*, which the society had published but which no one wanted to buy. This was, if nothing else, at least an ingenious way to remainder a dud. But the society's impecuniousness threatened a greater harm than merely filling Edmund Halley's library with unwanted copies of a boring treatise: they could no longer afford to publish Newton's *Principia*. The solution? The clerk would pay for it himself.

Still, it wasn't enough that Halley beat the work out of Newton, and then even paid for it. There had to be another hurdle thrown up in front of the intrepid clerk.

The *Principia*

The full title of the masterpiece was *Philosophiae Naturalis Principia Mathematica* (The Mathematical Principles of Natural Philosophy). It was to be in three books. Two of them were already in Halley's hands. But meanwhile, Newton learned that Robert Hooke was claiming to have been the originator of some of the ideas laid out in the *Principia*. Digging in his heels in the style of that truculent day, Newton wrote Halley of the last book: ". . . the third I now designe to suppress. Philosophy is such an impertinently litigious Lady that a man had as good be engaged in Law suits as have to do with her. . . ."

Diplomat and midwife both, Halley loaded up his supply of soothing words and set about prizing the last volume out of Newton's offended hands, writing, "Sir I must now again beg you, not to let your resentments run so high, as to deprive us of your third book. . . ." The third book was essential to the work, and while there is no doubt that Halley was motivated by the highest ideals in supporting Newton and contriving to comfort him, he can only have contemplated grimly the prospect of his house filling up with unsalable copies of the incomplete *Principia,* squeezed into whatever space was left over from *The History of Fishes*.

What did Halley accomplish when he finally managed to beg and flatter and cajole the manuscript into print? Nothing less than scientific revolution. Were it not for Halley, the *Principia* of Sir Isaac Newton would simply never have been born. That "divine Treatise" that drew together in a rigorous system the motions of the heavens and the ordered scheme of things on Earth would have moldered among the clutter of papers in a study at Trinity College, Cambridge.

Deep-Sea Diver

Though it was costing him money to stay at the job, the clerkship of the Royal Society suited Halley's fervid mind perfectly. He delved into everything, from a method for keeping flounders alive in baskets immersed in water so they would be fresh when sold, to a project for establishing the exact date that Julius Caesar landed in Britain. And then there was deep-sea diving.

In 1691, Halley began to experiment with a diving bell. The work went on aboard a naval frigate off the little town of Pagham on the Sussex coast. At first, the experiments were unsuccessful. But Halley was not put off, and later he himself made a descent in one of the huge, unwieldy contraptions.

The bell was a massive, cone-shaped device, three feet across at the top, five feet across at the bottom, and five feet high. It was heavily weighted to drive it down into the sea against the lift of the air trapped within its cavity. There was a valve on top of the cone to let out stale air. Fresh air was lowered in casks weighted with bands of lead. These casks were drawn alongside the submerged bell, unstopped, and their contents allowed to bubble up inside the chamber. It proved to be necessary to change the air in the bell every fifteen feet of descent.

A narrow bench ran around the inside of the chamber, and it is warming to imagine the animated and curious clerk of the Royal Society, so used to watching the sky, perched in this damp, black interior, no doubt happily wigless, and descending into the realm of the fishes he had had such a generous opportunity to get to know. According to Halley, the pressure on the divers' ears was painful, and became more painful as the bell was lowered deeper into the water. But he says that eventually the ears managed to equalize, and he recommended the application of almond oil to speed this process. No detail was beneath Edmund Halley, and the image of the great scientist swaying down into the murky coastal waters of England is a vivid one.

Ultimately the trials were a great success, and Halley was able to report to the Royal Society that he had kept three men at a depth of ten fathoms (sixty feet) for one hour and forty-five minutes "without any of the least inconvenience and in as perfect freedom to act as if they had been above." The claim is a little extravagant. This stuff was not exactly scuba gear.

The Bright and Restless Gaze

As a mathematician, Halley was fascinated by mortality calculations, and even provided some estimates of human longevity for insurance companies to use in working out what an annuity would likely cost them. He also had stern words for those who bemoaned the brevity of life:

How unjustly we repine at the shortness of our lives, and think ourselves wronged if we attain not old age. Whereas it appears hereby, that the one half of those that are born, are dead in seventeen years time. So that instead of murmuring at what we call untimely death, we ought with patience and unconcern to submit to that dissolution, which is the necessary condition of our perishable materials, and of our nice and frail structure and composition: and to account it a blessing, that we have survived perhaps many years that period of life, whereat the one half of the race of mankind does not arrive.

The ingenuity and freshness that Halley wielded whenever he turned his penetrating gaze upon some new problem are as fascinating to observe at work on the smaller puzzles as on the greater. Searching for a way to determine the extent of an area of land, he hit upon a solution almost hilarious in its simplicity. He would cut from a map the piece of paper that represented an area he already knew, and weigh it. Then, selecting the area he wished to determine, he snipped it out of the map and weighed it against the original piece. Let's say the first piece was one ounce, and Halley knew the area it represented was 100 square miles. Then if the next piece were two ounces, he reasoned that the area represented was 200 square miles. It was surprisingly accurate. Using this technique, he calculated the area of England at 38,660,000 acres. The actual acreage is 34,475,000. The error is barely more than 3 percent.

Immortality

It was not until 1695 that Halley began the awesome task of computing the return of the comet that would forever weld his name to one of the most familiar objects in the whole of the Solar System. It is ironic, too, that a simple lump of frozen water should eclipse so utterly the rest of his life's work.

Tycho Brahe had established that comets belonged to the upper atmosphere. But little was known about them. They were difficult phenomena for those early astronomers to study, since they arrived and vanished so swiftly, often beyond the reach of their telescopes. Newton's *Principia* placed the comets firmly within the grasp of the Sun, but Newton believed their paths described parabolas.

Using Newton's new calculus, Halley set to work. In an age when tiny, hand-sized calculators are used for even the simplest arithmetical functions, it is staggering to contemplate the task that confronted Halley. He would have filled hundreds of pages with the arduous, exacting business of computation. As well as the attraction of the Sun's gravity, drawing the comet along its orbit toward the center of the Solar System, he would have had to calculate the precise orbit of each planet, reckoning the effect of the gravitation of such giants as

Jupiter and Saturn on the flight of the passing comet. It was a mighty labor, and Halley concluded from it, correctly, that the path described by comets was an ellipse.

Seizing upon the comet of 1682, Halley determined an orbit of seventy-six years. Knowing it was unlikely he would survive to see it himself, he bid his contemporaries to watch for it, and if it should appear as he predicted "to acknowledge that this [the predicted return of a comet] was first observed by an Englishman."

This would be a tidy place to end the chronicle of Edmund Halley, packing him off into the countryside with his telescope, there to spend his last decades roasting chestnuts over gently flaming piles of *The History of Fishes*. But Halley was not yet forty, and had several more careers left within his energetic frame.

The Mint and the Czar

In 1696, Isaac Newton, then warden of the Royal Mint (these early scientists were nothing if not flexible), got Halley the position of deputy comptroller of the mint at Chester. The mint was then engaged in the monstrous job of recalling all the conventionally designed silver coins in the kingdom and replacing them with coins bearing milled edges. A milled edge carries scores of tiny grooves, like the edge of the American dime. This huge undertaking was aimed at curtailing the practice of clipping bits of silver from the edges of coins, by which many prosperous scoundrels were enriching themselves at the expense of the realm.

While in Chester, which he hated and considered a wasteland, Halley kept his friends back in London informed about any curiosities that passed before his quizzical eyes, like the report of the birth of a puppy to a *male* greyhound. It must have been a great relief to Halley to get free finally of Chester, where he had become embroiled in a bitter fight with two inferiors whom he suspected of turning a profit for themselves as the silver passed through their hands. Halley's superior at Chester defended the subordinates, likely because some of the purloined coin was clinking neatly into his own pocket. When at last he left the dismal city behind him to return to London, there were duties awaiting Halley exhilarating enough to brighten the cheeks of any man.

The Russian czar, Peter I, had come to England to study shipbuilding in the yards of the master wizards of that complex craft, the East India Company. With his retinue, the young autocrat was encamped at Sayes Court, a country house near Deptford. Peter was a rarity among monarchs, a man of intense curiosity who would stoop to almost any task in order to master it and more fully comprehend the craft of which it was a part. While visiting the company's yards at Deptford, the sovereign of all the Russias would cheerfully shed his coat and shirt, grasp a planing tool, and plunge himself into the manual niceties of the art he had come to study. By mastering these skills himself, Peter intended to gather into the very calluses of his hands the knowledge he sought to bring to his backward empire.

Brutal and wild, the emperor who would bludgeon his people into the modern age and campaign his way into history as Peter the Great was nonetheless a man whose respect for knowledge was consuming. From the very beginning, he and Halley, who traveled to Deptford to tutor the ruler in science, found one another kindred spirits. While there is no evidence that Halley took part in any of the spectacular orgies of drunkenness and destruction that left Sayes Court a ruin, he certainly dined with the czar, and his own appetite for drink was reportedly keen.

Mutiny at Sea

Shortly after Czar Peter finally staggered aboard the vessel that would carry him away from England (much to the relief of the appalled owner of Sayes Court), Halley boarded a ship of his own—this time in command. The king had appointed him captain of the Royal Navy's *Paramour,* a small craft known as a pink. Originally of Dutch design, pinks were well suited to the work of exploration that Halley planned to execute. They were sturdy, had a shallow draft, and were broad in the beam, with plenty of room to store supplies and equipment.

Paramour was fifty-two feet long, with a beam of eighteen feet and a draft of seven and a half feet. She displaced eighty-nine tons. In fact, she was a good deal smaller, and certainly a lot less comfortable, than some of the fat yachts that float so placidly at their moorings today in any of the pricier anchorages of Florida. And yet Halley was planning to take her to sea for a year on a voyage carrying them to West Africa, eastern South America, and the ". . . terra Incognita, supposed to lye between Magellan's Streights and the Cape of Good Hope." This was, of course, Australia. On the way home, the plan called for knocking off the East and West Indies, as lightly as if it were a pop around the corner for a quart of milk. And every foot of the way they were to chart and map and sound. It was hopelessly ambitious, and the first voyage of its kind ever attempted by the British government . . . *seventy years before Captain Cook!*

Right from the start Halley had trouble with the crew, who resented the command of a landsman and pretended not to understand his orders. In spite of the willful stupidity of the men, Halley managed to chart the waters off the Cape Verde Islands and to map the coast of Brazil. By then, the behavior of even the officers had become so refractory that the besieged scientist set his ship for Barbados, hoping to find a naval flagship in the roadstead and exchange his troublesome company for more agreeable sailors. There was no British force at the island, and in May of 1699, in despair, Halley set course for England. By June, he could bear the conduct of the first lieutenant, his senior officer, no longer, and he clapped him under arrest and held him in confinement for the rest of the voyage "and brought the Shipp well home from ner the banks of Newfoundland without the least assistance from him." If the lieutenant had imagined that a master of astronomy like Halley could not figure out how to bring a ship safely home across the ocean, he was very much mistaken.

It turned out that the lieutenant, one Harrison, had tried to have the Royal Society publish a method of his invention for determining longitude at sea, and Halley had been one of the members who had panned the manuscript. More than that, Halley suspected that Harrison, a career sailor, resented him because he couldn't use "the whole Sea Dictionary as perfect as he." Anyone who has ever tried to use plain English to talk to a sailor will know how Halley felt.

The Second Voyage

In the fall of the same year, Halley took *Paramour* to sea again, and this time sailed the little pink right down to the bottom of the world. In the freezing seas around the Falklands and South Georgia Island, the men saw "a couple of Animals which some supposed to be Seals but are not so; they bent their Tayles into a sort of Bow . . . and being disturbed show'd very large Finns as big as those of a Large Shirk The head not much unlike a Turtles." Probably what they saw were killer whales. At this time, Halley's cabin thermometer registered only four degrees above freezing, and *Paramour* began to meet icebergs. After one particularly frightening encounter, when the little vessel was surrounded by mountains of ice that Halley feared would stave her in, he rounded to the north and sailed for the sun.

Twice on this voyage, Halley was taken for a pirate. The first time he was actually arrested in a Brazilian port at the instigation of a British consul who doubted his papers. The second time he found himself standing on his deck with cannonballs slapping through the canvas and past his ears as one of his fellow citizens, with the means to do something about it, took him for a pirate again. But Halley had not survived the czar of Russia to die by cannonade, and the expedition produced a set of fine charts of the Atlantic that guided mariners for decades.

Honors at Home

In 1703 Halley was at the height of his reputation and influence when the Savilian Chair of Geometry fell vacant at Oxford. The astronomer royal was now an enemy of Halley's after some disagreement about the measurement of tides in Ireland. Thus Flamsteed wrote: "Dr. Wallis is dead. Mr. Halley expects his place. He now talks, swears, and drinks brandy like a sea-captain, so that I much fear his own ill-behavior will deprive him of the vacancy." But a little swearing and swigging was not going to bother Halley's admirers, and he got the job anyway.

So compelling are Halley's achievements that it is easy to forget how long ago he lived. Observing the phenomena of nebulae, he concluded they were made of "Light coming from an extraordinary great space in the Ether." We now know that these nebulae are vast fields of gas, flowing from the stellar radiation born in the furnaces deep within their measureless precincts. Even to mark these nebulae was an impressive achievement. To have postulated so nearly their nature is astounding.

Studying the aurorae, our northern lights, he theorized that the luminous display originated in the earth, which exuded some substance controlled by magnetism. His hunch on the connection of magnetism to the aurorae was right, although present theory has the pulsing, silky light pouring into our atmosphere from above.

The intractable and venomous astronomer royal, John Flamsteed, died in 1719, and Edmund Halley ascended the throne of his career. But when he arrived in Greenwich to assume his post, the Royal Observatory was an empty shell. All of the instruments had been Flamsteed's personal property, and when he died his widow stripped the rooms bare. Halley had to wrest 500 pounds from the board of ordnance to fit the place out. Ten years later the royal consort, Queen Caroline, visited the famous astronomer at Greenwich. Finding that Halley had at one time served the Crown in command of a ship, the doughty pink *Paramour,* Her Majesty ordered that he receive the half-pay that was the right of any retired naval captain. This would have been welcome money indeed, and must have eased his last years.

On January 14, 1742, in his eighty-sixth year, the astronomer royal, feeling tired, "asked for a glass of wine, and having drunk it presently expired as he sat in his chair without a groan."

CHAPTER 5

Death of the Dragons

Now all the dragons are dead, but once this planet belonged to them. Some people believe they were murdered; others, that they died a slow and choking death, their very home having become poisonous to them. However the great beasts fell, nothing is left to mark their dominion but a few footprints, some old bones, and the sounds of debate as paleontologists argue about the manner of their passing.

It is 136 million years ago, and we are watching from above in our time balloon as the day begins along the shore of a forested lagoon. The great boiling Sun has already risen above the steaming forests, and the wild colors of the Cretaceous dawn are splashed into the sky around us. There is a quick, dark flash in the air beneath, and a small, batlike lizard cuts in a sharp turn and seizes a fat and blundering moth in its little teeth. There are other creatures flying, too. A pair of fanged lizard-birds the size of herring gulls patrols the surface of the lagoon, gliding gull-like along the water, waiting for some primitive fish to venture near the surface. When they spot prey they brake suddenly in an upward swoop, turn, and dive into the water, snatching the fish and flapping awkwardly free of the water. Climbing laboriously, one of the heavy birds rises to an altitude of about thirty feet, flips the fish up into the air above with a jerk of his long head, opens his beak wide, and catches and swallows the dropping fish. In a slow and circling descent the lizard-bird returns to the surface and resumes his patrol, back and forth across the waters of the lagoon.

We are so riveted by the action beneath us that we have failed to hear the splashing and slurping sounds that have grown over the last few minutes. Now they are loud enough to draw our attention, and we turn to look along the shore of the lagoon, seaward. Around a point of land a small reptilian head appears, suspended in midair at the end of a thick neck. Slowly the head sinks, slurps and sucks about in the water, then rises with a mouthful of vegetation, chewing placidly. Each minute, a few more feet of this midair neck appear, and still we have not seen the body to which it is attached. Three of these heads have appeared around the point, trailing their thick necks, before the body of the first heaves into view. Slowly a herd of these vast sauropods —immense vegetarian dinosaurs with elephantine bodies and long necks and tails—appears. There are almost thirty great beasts noisily slogging through the water, their feet sucking as they pull them from the muddy bottom. Several of the largest in the herd are eighty-five feet long, their bodies arching up to a height of twenty feet at the hips. This is Brachiosaurus, weighing in at seventy-five tons. Brachiosaurus moves along at about two miles an hour in his plodding way. His brain is almost inconceivably small for so large an animal, but all it has to do is tell his feet to move, his head to dip, his mouth to open and close, and his head to rise again. This is not an elaborate program.

Change in Eden

This is the late, or Upper, Cretaceous period, the final period of the Mesozoic era, the era before our own. Man inhabits the later Cenozoic era, the era that has seen the rapid evolution of mammals and birds, of grass and shrubs, of most of the flowering plants, in short, of much of what we as humans consider beautiful and natural in our world. But our own period is a scant 2 million years old. Dinosaurs, on the other hand, appeared in the late Triassic and lasted until the end of the Cretaceous or the beginning of the Tertiary era—about 100 million years. Obviously, they were very successful animals to have lasted so long. Why did they die, instead of hanging around to eat anything troublesome that showed up, like man?

By the end of the Cretaceous period, much was happening on Earth that would profoundly affect its future life. The continents had been slowly dividing since the Triassic period. The supercontinent Pangea had already broken up: North America had drawn away from South America, South America from Africa. India cast off Australia, an act it probably regrets. Between these landmasses, the waters of the new oceans fingered their way. In the Northern Hemisphere particularly, there was an explosion of new flora and fauna. Before this evolutionary burst, the ginkgo was the only broad-leafed tree on the planet. It is with us still, an ancient tree, the primitive veining of its leaves betraying its antiquity. There is none of the delicate tracery of the leaves we see today. The veins fan out in simple, hardy symmetry. The limbs of the ginkgo are stout and do not fork. Instead, each branch grows short twigs, and these twigs bear the fan-shaped, cloven leaves. No other tree resembles this immemorially old survivor, a living fossil.

Now the other deciduous trees appeared: oak and maple, poplar and walnut, hickory, magnolia, viburnum. There were still slender land links between Asia and North America, North America and Greenland, Greenland and Europe. Across these links an energetic evolutionary traffic honked and beeped its way around the Northern Hemisphere. The rich bonefields of Montana and Alberta tell us there was a multiplicity of animals. Small mammals, probably rodents, had appeared in the Cretaceous. There were a few birds we would recognize today, owls perhaps, to supplement the population of flying lizards and pterosaurs. Placid duck-billed dinosaurs roamed the rich forests and the beautiful plains. Up in the higher lands the Pachycephalosaurs appeared. These were dome-headed dinosaurs, living in herds. The structures of those herds, we now know, resembled the herd structures of the sturdy mountain sheep and goats that populate the upland meadows of the Rockies today.

And then this world vanished.

Tentative Autopsy

We know there were no dinosaurs left on Earth after the end of the Cretaceous period because there are no fossils of them in any of the rocks deposited after the Cretaceous. The dinosaurs had ruled Earth for 140 million years, dominating our remote mammalian ancestors, who replaced them. These early mammals were a puny lot, mostly insectivores, and it is strange indeed to ponder the sight of the fabulous dragons tumbling from their thrones atop the food chain, the Tyrannosaurs, the rhinoceroslike, two-ton, armor-plated Monoclonii, to be replaced by . . . mice? Here is some of the orthodox thinking that surrounds this paleontological whodunit.

The new flora that appeared, the flowering angiosperms, may have contained more toxins and alkaloids than the dinosaurs could bear. It may be that the dinosaurs had such primitive taste buds that they simply munched away at whatever they found, finally keeling over with food poisoning from too much of a bad thing. Or the new plants appearing all over Earth could have interfered with the reproductive process of the great animals, causing them to lay infertile eggs.

Another theory suggests that a dietary calcium deficiency made the dinosaurs' eggs thin-shelled and affected the development of the embryonic hatchlings. This same phenomenon is at work today in the plight of some eagles and falcons. Egg malformation caused by some dietary anomaly can threaten the survival of modern species as well. Perhaps lack of calcium did to the dinosaurs what DDT may do to some of our own predators.

And there was the cold. By the end of the Cretaceous, cooler conditions had settled over much of the globe. With the drift of the continents and the erection of new mountain barriers, many of the dinosaurs could no longer migrate to escape the cold. Their skins were scaly and had no effective insulation. With the drop in their body temperatures, say biologists, they would simply have become unable even to eat. In short, they were paralyzed by cold.

And yet other inhabitants of the day survived: turtles and crocodiles, snakes and lizards. Why? Perhaps they ate different foods than the dinosaurs, and their ability to hibernate saved them from the cold. When the larger plant-eaters died, the great carnosaurs [carnivorous dinosaurs] that fed on them would also die, foraging through their cold and desolate world in futile search for the food they would never find.

Those are a few of the theories. But there's another one—more sudden and more vicious and more final. The dinosaurs were slaughtered by a comet.

Deathblow

Imagine the dinosaurs lumbering about in their steamy Eden. Everything they need to eat is readily at hand: plants, the fishes in the teeming lagoon, and one another. There is really nothing in the way of hard work for their diminutive brains to tackle.

Suddenly the air is shattered by a terrific sonic boom, followed by a rising scream growing to a pitch unbearable even to the ears of a laid-back lizard. The whole atmosphere lights up, and the light grows in intensity, blooming and burning across the sky, everything lit in the awful glare that pours through the world. The wailing shriek screams on above the planet.

Then, impact.

For the dinosaurs in the immediate vicinity—say, within 1,000 miles of the impact—death would be quick. Punishing shock waves rushing out from the epicenter would burst their eardrums and pop the eyes out of their heads. The blow of the shock waves would cut the mighty beasts down as in a murderous fire. For those not simply massacred in the hideous wreckage spread by the impact of the comet, death would still come swiftly within the 1,000-mile radius. The force of the comet's impact would create an explosion 100 million times more brutal than the explosion that devastated Hiroshima. Thick, choking clouds of dust and debris would boil up into the sky. A deadly, lung-coating smoke of vaporized material would mushroom out from the epicenter. Great beasts staggering about in the awful dust, blinded, some of them deaf, roaring in pain and anger and confusion . . . the scene would have been a scene from hell. Maddened, the animals attack anything they bump into, tearing at and striking the dim shapes of trees. Some die in the sightless carnage, the victims of slashing claws and teeth. Others lie on the ground choking out their lives, asphyxiated.

That is what happened to the animals in relative proximity to the horrible blast. Death came more slowly to the rest. Relentlessly, steadily, the wind spread the pall around the globe. At first, the sky appeared only overcast. As the days passed and the currents shifted high above the surface of the land, the great miles-high cloud of pulverized rock and dirt shifted and coalesced, adhering finally into a uniform, opaque envelope of debris wrapping Earth like another atmosphere, an atmosphere of death. Who knows how long it took to kill the rulers of Earth? Probably some survived for years. These had a reprieve. But they were not spared.

Finally the plants died, starved of sunlight. With the plants went the herbivores. With the herbivores went those kingly butchers, the ruling carnosaurs. Who was left? Our great-great-great-great-great-great-great-etc. grandpas and grandmas, those little mammals who topped up their tanks with insects and worms. A few of the predators who made their meals out of our ancestors survived, too. Some marine fauna scraped by on what was left in the dying sea after the obliteration of the phytoplankton, the microscopic foundation of the food chain, itself dead from the unending dark. This paltry remnant of the rich animal population that had once thrived on the bountiful planet, this remnant scraped and scrabbled along until, at last, decades or centuries later, the fearful cloud blew away and Earth began to soak up life once more.

This is the theory proposed by a team of scientists from the University of California at Berkeley: Walter Alvarez, a geologist; his father, Luis Alvarez, the Nobel laureate in physics in 1968; and two nuclear chemists, Frank Asaro and Helen Michel. They were not looking for dinosaur killers when they set out on their researches. They were searching for a way to measure the rate at which certain sediments are laid down on Earth.

The Iridium Clue

The key to this compelling saga is an arcane element called iridium. Earth's supply of iridium is taken aboard fairly regularly through a steady spattering of the stuff from space. Space has its own supply somewhere else in the universe, and never seems to run out. If the scientists could measure the amount of iridium in a given deposit of rock, then they ought to be able to tell how long that deposit took to form, since the rate of fall of iridium is known. What they found astonished them, and the discovery has set the world of speculative paleontology, a more rough-and-tumble place than you might expect, buzzing with controversy.

At Gubbio, Italy, they found a layer of clay separating the Cretaceous from the beginning of the Tertiary, and containing thirty times the iridium that was present in geological formations on either side of that time level. At Stevns Klint, Denmark, they found deposits from the same period, the Cretaceous/Tertiary with *160 times more than the average iridium level before or after*. Similar strata high in iridium have been located in New Zealand and Spain. To restate the phenomenon, a substantial layer of iridium, as much as *16,000* percent heavier than usual, was being laid down on the surface of Earth at precisely the time that dinosaurs were laying their heads in the dust for good.

The scientists believe the freight train that smashed into Earth with this load of iridium was an asteroid, orbiting the Sun in an unruly belt of brothers between Mars and Jupiter. Every now and then one of these unprepossessing planetesimals nudges another, and the nudgee comes weaving silently out of the blackness, headed for no good. The team proposes that the asteroid was a rock the size of Manhattan, and that it slammed into Earth at 50,000 miles an hour, gouging out a crater the size of Connecticut. If you were a dinosaur, this was the time to scribble a note to posterity and stick it in a bottle, because tomorrow—she never come.

Many scientists shy away from these catastrophic explanations. And indeed, there does seem to be a pretty brisk trade in disaster on the part of those who market terror scenarios. But just because a lot of people make money by saying that the universe is a very violent place does not mean that it is really peaceful. It is not. And as Paul Hoffman, a science writer and physicist, has noted in "Asteroid on Trial" in *Science Digest*, June 1982:

. . . extraterrestrial forces brought the Earth into existence, and it is insular to think that, having set the Earth in motion, they will never tamper with it again.

The asteroid theory is attractive because of the conspicuous failings of the many down-to-Earth theories of dinosaur extinction that have been put forward in the past two centuries. Although little is known about the declining years of the dinosaurs, the beasts are so engaging, and their disappearance so mysterious, that the paucity of evidence has not stopped palaeontologists from trying to get an indictment against a culprit.

One popular but romantic notion is that man's ancient mammalian relatives, tree-climbing, insect-eating, shrewlike creatures, paved the way for our existence by scurrying from one dinosaur nest to another sucking and devouring the eggs. The theory won't hold water because the mammals were too small and too scarce to have demolished that many fetal dinosaurs. Moreover, for 100 million years the dinosaurs and mammals commingled without incident. If any generalization can be made about their coexistence, it is that the huge, lumbering reptiles managed to suppress the rise of the small, active mammals, not the other way around. The ratio of an animal's brain weight to its body weight is a measure of its intelligence. By this measure some of the meat-eating dinosaurs in the Late Cretaceous, such as the sprightly, long-legged Stenonychosaurus inequalus, were as smart as the first mammals. Only after such brainy dinosaurs vanished did the mammals develop along their many evolutionary paths, including the one that led to Man.

Constipation, Tektites, Supernova

There is another problem attached to attempts to explain the disappearance of the dinosaurs by exclusive reference to terrestrial events. Along with the dinosaurs perished a great many other forms of life, plant and animal both. Any model for dinosaur extinction, therefore, has to include these other victims. It seems unlikely that constipation due to changes in diet, which is one theory, could have eliminated all those animals. The vision of the Mesozoic era drawing to an undignified close amid the groans of constipated dinosaurs invites a skepticism of, well, mammoth proportions.

For the catastrophists there are four possible killers of the dinosaurs. A comet and an asteroid are two. The others are a cloud of microtektites and a supernova.

The microtektite theory is the brainchild of John O'Keefe, who was, at the time he proposed it, a geophysicist at the Goddard Space Flight Center. Tektites are little beads or stones of a glassy texture. They contain rich loads of iridium. These little tektites are beautiful objects that come in an enchanting array of shapes, sometimes resembling teardrops, sometimes rods or disks, sometimes even tiny dumbbells. O'Keefe's theory was that massive volumes of these microtektites were blown into the atmosphere from volcanic eruptions on the Moon. Many of these tektites would simply be grabbed by Earth's gravity and drawn onto the surface of the planet. If there had been a period of very heavy volcanic activity on the lunar surface, that might account for the layer of iridium discovered by the Berkeley team: many tektites; much iridium. But not all of the little glassy beads fell to Earth. Some remained aloft, forming a cloud circling Earth in the equatorial plane. From space this band of tektites might have looked like one of Saturn's rings. From Earth it would not look nearly so fetching. The effect would have been to block the sunlight, starting the process that led to the death of plants and animals alike.

If this theory appeals to you, by all means climb on the bandwagon; you will find a lot of room. The scientific community generally has trouble accepting the tektite model, although no one has managed to refute it.

What about a supernova? Terms like *supernova* are bandied about as lightly today as if we were talking about firecrackers. We are not. A stellar explosion of sufficient magnitude to be classed as a supernova can release as much radiation as would be produced by 1,000 billion billion ten-megaton bombs. That is an explosive force of 10,000,000,000,000,000,000,000,000,-000 tons of TNT. A supernova pours into space for two weeks as much energy as an entire galaxy puts out in the same time. However, even this great an explosion relatively nearby would not have been enough to lay down the amount of iridium found in the ancient clays at the edge of the Cretaceous/Tertiary. In 1981, Wallace Tucker, then senior theoretician at the Harvard-Smithsonian Center for Astrophysics, suggested that what the supernova theory needed was a *super*-supernova,

. . . an explosion 1,000 times larger than a supernova. Such an event could have occurred a thousand light years from our solar system and still have caused mass extinctions on Earth through radiation sickness and the climatic effects of the dramatically increased background radiation. The force of the explosion would have evacuated an enormous volume of space, pushing ahead of it a dense sheath of iridium-enriched gas. The gas would have slowly rained down on the Earth as the shell swept past the planet thousands of years later.

In fact, such an event might actually have happened. Surrounding the solar system, and hence the earth, is a vast ring of gas and dust. About 4,000 light years in diameter, the ring is expanding at the rate of 10,000 miles an hour. Just inside this ring is a belt of young stars, each of which is less than 60 million years old. From the current dimensions of the ring and its explosion velocity, we can estimate when the ring first started expanding. The answer: 65 million years ago, just when the mass killing occurred.

At that time a titanic explosion apparently occurred about 800 light years from Earth. As the shock wave from the explosion roared through interstellar space, it swept up interstellar matter into a dense shell. Clouds of gas, dust that got in the way, were compressed, triggering the formation of the new stars that now appear at the edge of the ring.

But there is a problem with this theory. Several heavy elements are created deep within a star that is on the verge of a supernova explosion. Iridium is one of these elements, and so is plutonium 244. When the star blows up, these two elements, along with everything else, are driven into space. Theoretically, the ratio of plutonium 244 to iridium should remain a constant. In other words, it is possible to know how much plutonium 244 should be in those iridium deposits if the iridium came from an exploding star. When they went looking for it, however, the scientists were unable to find *any* plutonium in the iridium layer.

Good-bye, supernova.

Back to the Comet

A team of researchers at the Massachusetts Institute of Technology has developed a theory it believes accounts for many of the anomalies unexplained in other models put forward to account for the mass extinction. One of these anomalies is the fact that, whatever happened back there, it killed off more life in the oceans than it did on land. More puzzling than that, the cataclysm almost exterminated marine organisms with shells made of calcium carbonate, yet spared those with shells of silica. It hammered the mighty ruling reptiles into the ground, but spared the insignificant little mammals who moved in to take over.

Here's how they say it all happened.

A large comet comes belting into Earth's atmosphere at sixty times the speed of sound, or a little better than 65,000 feet per second. Blazing through the terrestrial atmosphere at this terrific speed, the comet builds up a powerful shock wave, effectively transferring energy from itself to the surrounding air. This is no outsize monster, mind you, just an ordinary-sized comet, like Halley's. The effect would be immediate.

A great bulge of superheated air follows the comet Earthward. Within this bulge the air is heated to a temperature of several thousand degrees Celsius. This streaking mass of air causes certain chemical reactions to take place in the atmosphere through which it passes, creating compounds that under normal circumstances are very rare at that level. One of these compounds is nitric oxide, which oxidizes to form nitrogen dioxide. The nitrogen dioxide mixes with the water present in the air to form nitrous acid. This is exactly the same chemical we call acid rain. While it is true that nitrous acid is not a particularly strong acid, it is nevertheless toxic and can poison a wide range of organisms. One of the MIT researchers compares its action to that of sodium nitrite, a compound in use as a food preservative, which works by poisoning bacteria living in that food.

According to the researchers, the amount of toxic nitrous acid created would be a function of the amount of energy transmitted into the surrounding air by the rushing comet. For every erg of energy created by the comet, a billion molecules of nitric oxide—the compound that starts the chain reaction—would be created. (An erg is the energy equivalent of accelerating one gram of mass to a speed of one centimeter per second per second.) But that is a pretty clinical way to describe what happens when a comet pounds into the air shell wrapping a planet like ours. Try this:

The amount of energy released by such an unwelcome visitor would be equal to the explosion of 250 trillion (250,000,000,000,000) tons of TNT. This is about 4 billion times greater than the force of the bomb dropped on Nagasaki. It would create about 100, 000,-000, 000, 000, 000, 000, 000, 000, 000, 000, 000, 000,-000 molecules of nitric oxide. That is 100 thousand billion billion billion billion molecules of nitrous acid.

Since two-thirds of the surface of Earth is covered by water, the MIT team reasons that the comet that brought this unpleasantness about probably plunged into the sea. The explosion would have been stupendous. As the comet hit, it instantly burst into a cloud of gas, vaporizing a great volume of seawater along with its own mass. A tremendous volume of hot water, mineral salts, and particles of the former comet were all blasted high into the air by the explosion. Most of this mass simply fell back to Earth, but some of it, the deadliest, stayed aloft. This airborne detritus formed into hydrochloric acid and sulfur dioxide, two of the meanest chemicals a gentle planet like ours ever had to face.

The Burning Rain

Today, we are responsible for these plagues ourselves. Coal-burning factories and generators spew forth a pall of such toxicity that hundreds, maybe thousands, of wilderness lakes in the northeastern United States and Canada are sterile and lifeless. We are not alone. In China, where progress has been maintained at a heavy environmental cost, the authorities have only recently recognized the breadth of the devastation wrought by their uncontrolled production of unadulterated poisons as an accepted by-product of industrial expansion.

Sixty-five million years ago, some scientists conjecture, the same kind of pollution began to fall from the skies—but in quantities appallingly higher—an unrelenting, searing, killing rain of acid. The rich, lush world of the dinosaurs, whose empire had lasted for 100 million years, began to burn in the scorching rain. On land, the acid rain defoliated huge tracts of forest, condemning the herbivores to starvation. Every day, all day, it fell upon the desperate animals, burning their eyes, stinging their tongues when they ate, burning their throats when they drank from the little pools that formed in the hollows of the rock. The little mammals who were to inherit the ruined planet from its dying nobility—how did they survive? We don't know. Perhaps they hoarded seeds for food, and drank water that condensed on the undersides of remnants of leaves or in caves in the rock. Their needs were more modest than were those of the fallen knights of the land. They survived.

In the oceans, the killing would have gone on as methodically as on land. The tiny microorganisms, the phytoplankton, which dwelled near the surface of the ocean, would die in the acid. With the phytoplankton would go the whole marine food chain. The plankton-eaters would die next, and then the fish that preyed on them. Acid rain would dissolve the protective casings of the animals whose shells were made of calcium carbonate. Those with silica shells survived. The silica was impervious to acid, as is glass, which is made from it.

That's one way they might all have died. Or it could have happened like this.

The Monstrous Wave

The comet slapped into the ocean so hard that it raised a tidal wave three miles high. This wave, half as high again as a thousand-story building, the equivalent of fifteen Empire State Buildings stacked one atop another, would do an effective job of scrubbing Earth clean of a lot that goes clomp in the night. This theory is tendered by two scientists at the California Institute of Technology. The Caltech scientists used a series of computer studies of the effects of meteorite impacts on Earth to develop their model of the likely effects of the much harder collision of a comet.

The theory begins with the assumption that a comet like Halley's banging into Earth would leave a crater 100 miles in diameter. If this crater were on land, we could find it, and tell from the geology how old it was. So far, we have not found such a crater. But if the comet had plunged into the ocean, the scar could easily have disappeared. Something like 51 percent of the ocean floor that existed at the time of the dinosaurs' death has disappeared into the interior of Earth through a process called subduction. So the mere inability to find such a crater does not mean that one never existed.

The Caltech team fed into their computer an object seven miles across. It would take an extraterrestrial object of at least that size, they reasoned, to lay down the amount of iridium found in the 65-million-year-old deposits at the edge of the Cretaceous/Tertiary. At that size, the comet would have exploded with the force of a 1-billion-megaton bomb, or 1,000 million million tons of TNT. This is what the scientists imagine:

- At twenty-five miles per second the comet tears out of the sky and slaps into an ocean that holds a depth of water of at least four miles.
- In one second, a huge cavity twenty-five miles across and twenty miles deep appears.
- A few seconds later the entire cavity fills with superheated seawater and vaporized rock.
- Over the crater a bubble of steam thirty miles wide pushes up into the atmosphere to an altitude of twenty-five miles, completely covering and concealing the scalding crater beneath.
- Over a period of twenty-seven hours, heavy aquatic shakes emanate from the cavity, until a tidal wave three miles high envelops the whole of Earth.
- The wave sluices into the marshy lowlands, which paleontologists say formed the typical dinosaur habitat. Even if some of the beasts survive the onslaught of a weight of water three miles high, the receding wave would still strip the vegetation from the land and bury it in silt.

The Caltech scientists maintain that even if the comet struck dry land, the resulting earthquake, probably on the order of 12 on the Richter scale, would trigger a seismic wave at sea.

An ocean impact by itself would raise a terrestrial dust cloud containing as much as 100 times the mass of the comet. This great cloud would rise up into the sky, remaining there after the wave had sloshed away back into the ocean basins. To this darkening world, add heat. The scientists at Caltech believe that 15 percent of the energy generated by the comet's impact would go toward establishing an increase in temperature of 10 degrees Celsius in only a few hours.

Kenneth Hsu of the Swiss Federal Institute of Technology in Zurich thinks that just such a cometary impact may have poisoned the dinosaurs. According to Hsu, the load of cyanide carried by a comet would be enough to kill off the phytoplankton if it fell into the sea. With the extermination of the foundation of the marine food chain, global catastrophe would soon ensue.

Clockwork Killer

Even more menacing than the theories that attribute the violent end of the dinosaurs to the brutality of space are the claims that it will all happen again, presumably to us. There are scientists who believe that the kind of mass killings that drove the great reptiles into extinction are regular and predictable phenomena. Once every 26 million years, some awesome calamity befalls our otherwise peaceful little planet, and the life-forms that have evolved between calamities get their first real test. According to this theory, millions of species that have successfully established themselves on Earth between these catastrophes are erased from our planet forever. Now evidence suggests that the culprit comes from space.

The case for regular attacks from space first began to emerge in 1977. Two geologists from Princeton University, Alfred Fischer and Michael Arthur, became convinced that certain wholesale extinctions of sea life occurred in cycles. They came up with the figure of one slaughter every 32 million years. This is not the kind of guesswork that passes unnoticed in the world of science, and a paleontologist from the University of Chicago, John Sepkoski, Jr., was soon sniffing around in search of something more conclusive. Sepkoski put together records that chronicled the rise and fall of 3,500 different kinds of sea creatures. Analyzing this data with great care, Sepkoski and a colleague, David Raup, discovered twelve massive extinctions of sea life that fit into a cycle of one extinction every 26 million years.

Back at Princeton, geologists Arthur and Fischer took another look at their original figure of a 32-million-year cycle. This time they used more recent age-dating techniques, and came up with the same figure as their colleagues in Chicago: some awesome force was sweeping through the seas of our planet annihilating the marine life every 26 million years. But where did it come from? What could charge through the oceans with this awful killing ability? And how was it that the apparent slaughters arrived with such chilling regularity?

Most of the researchers studying the problem are able to accept that an exterminating force, like that which struck the world of the dinosaurs, might be a comet. *Once.* After all, such a death-dealer need not be all that monstrous in size. A comet only ten miles wide whistling in here at 100,000 miles an hour would explode with 10,000 times the force of the total of all the nuclear warheads on Earth. That is a pretty sharp blow to the body, even for a planet. It would certainly deliver the knockout.

Once.

But it would not come back every 26 million years to do the job all over again. And that is the principal problem for scientists who want to erect their extinction scenarios upon some violent extraterrestrial visitation.

Where does the regularity come from? What clockwork could tick through time to pelt us at predictable intervals with such savage alarms?

Shaking Loose the Comets

From Earth, the Milky Way appears to be a pale wisp of cloud, a strip of veil drawn across the top of the night sky. To the unaided human eye, the collected light of the Milky Way is blurred and indistinct. And yet it is the center of the Galaxy that we behold, where a hundred billion stars are born, blaze, and die in the seething galactic heart.

The Galaxy is a great, revolving disk, a flattened system containing perhaps 200 billion stars. From the dense core of the Galaxy—the Milky Way that we see in the night sky—spiral arms reach out to the more thinly populated edges of the mighty construct. Far out on one of these spiral arms dwells Earth. Our own Sun, a tiny star among 100,000 other stars, weaves slowly back and forth across the flattened plane of the Galaxy, trailing its fragile brood of planets along. Up it moves across the galactic plane, and back down again, following the spiraling motions of the arm that stretches out from the center of the Galaxy.

According to two scientists from NASA's Goddard Institute for Space Studies in New York City, our solar system makes an upward or downward journey once every 33 million years, with an uncertainty margin of 3 million years. Subtracting the uncertainty margin gives a possible figure of one journey every 30 million years. But even *that* figure could be off by a few million years. Putting a figure to when things happened such a long time ago is not like circling a date on the calendar.

Thus, it is possible to construct a theory in which mass extinctions on Earth coincide with the regular voyage of the Solar System up or down through the galactic plane of the spiral arms.

Like the seething center of the Galaxy, the spiral arms contain clouds of gas and dust from which new stars and planets will ultimately be formed. According to Michael Rampino and Richard Stothers, the two scientists from Goddard who have formulated the theory, the Solar System must pass through some of these clouds of dust or gas as it makes its way through the galactic plane where they are most heavily concentrated. As she sailed along through space with her attendant planets, the Sun's voyage into one of these mighty clouds of interstellar matter would be akin to a voyage into the heart of some savage storm.

Sailing into, or even *close* to, one of the dark clouds of interstellar dust, the Solar System would experience the tremendous gravitational forces attached to the cloud. While the cloud's gravity would not be enough to perturb the orbits of the planets, it would certainly be strong enough to claw at the 100 billion comets circling far beyond Neptune, crawling along at a snail's pace within the slowly shuffling herd of the Oort cloud. Encountering the terrific forces of the dust cloud, millions of comets would be simply shaken out of their patient orbits and come tumbling in toward the Sun . . . and Earth.

Some of the comets would be tearing along on orbits so eccentric and so far off that they would loop the Sun and whip off into space forever. Others would be shunted aside by the powerful gravitational fields that surround those two defensive linesmen of the Solar System, Jupiter and Saturn. Still others, too small to survive the journey, would be blasted into nothingness by the force of the solar wind as soon as they began to draw into the skies above the Sunward planets. But of course, not *every* comet has to strike earth in order to exterminate much of its life.

Only one is needed for doomsday.

When he was asked in early 1985 when the next crossing of the galactic plane would be undertaken by our solar leader, the Sun, Goddard's Dr. Rampino said: "It's there now." This is a chilling thought, but not one that ought to cause any immediate panic. For the great, dark clouds that lie along the galactic plane are not in a straight line. Some are above and some below the actual plane. This spreads the probability of plowing into one over several millions of years. It is this factor, too, that accounts for the necessity of building in a margin of error in calculating the years between celestial bombardments. And it is this random spreading of the clouds that accounts as well for different elapsed times between terrestrial extinctions.

According to Dr. Rampino, there was, for example, a mass extinction during the Miocene epoch, 11 million years ago. At that time, Dr. Rampino maintains—even *that* long ago—we were already drawing back toward the galactic plane on our *present* crossing, and were close enough to encounter a cloud. That is why scientists do not get too excited when their estimates differ by a few million years. There are simply too many random factors in space to permit the establishment of a firm time cycle.

If Earth has been savaged by comets in the past, then there ought to be evidence. And there is. Although Earth does not bear the record of all its recent collisions with extraterrestrial objects, as does the Moon, there are at least some craters around. Analyzing these craters, Rampino and Stothers conclude that Earth has been peppered with comets every 30 million years. The paleontologists at the University of Chicago who had studied the marine extinctions agreed that roughly 30 million years for a cycle between bombardments would also fit their own model.

And so the case is made for a regular hail of comets

shaken loose by the terrible forces of interstellar clouds as Earth trails along with the Sun on her awesome solar voyages through the galactic plane. But that is just one of the theories formulated to account for mass extinctions on Earth. There is another killer suggested.

She is called Nemesis.

Death Star

Most of the data marshaled to fit the hypothesis of a regular journey through the menacing galactic plane can also fit the stunning theory of a "death star," and in fact the death star theory conforms more easily to the 26-million-year figure originally suggested by the Chicago paleontologists. The death star theory has been proposed by a group of scientists at the University of California at Berkeley, and it is they who have named the killer star Nemesis.

According to the death star theory, Nemesis is a companion star to the Sun. A dimmer and much smaller star than its sister, Nemesis travels on an elongated ellipse so immense that it charges back into our own vicinity only once every 26 million years. Even now, astronomers are scouring the sky, examining many of the fainter stars that were ignored until the formulation of the death star model of destruction. The hope is that we will be able to identify Nemesis among the faint stars, and plot her menacing course. It was comets set in motion by Nemesis, the Berkeley scientists hold, that smacked into Earth 65 million years ago and set up the awful chain of devastation that wiped out the dinosaurs.

Howling into the Oort cloud at the fantastic velocity of an orbiting star, Nemesis would scatter a billion of the docile comets off in all directions. The death star would not have to actually collide with a comet in order to drive it from the comet cloud and set it falling in toward the Sun. The awesome force surrounding the passage of a star would be enough to rattle the comets loose, even from hundreds of millions of miles away, and the California scientists believe Nemesis could be farther out than that.

The Berkeley team—Richard Muller and Marc Davis of Berkeley and Piet Hut of the Institute for Advanced Studies in Princeton—believe that Nemesis is only about one-tenth the mass of the Sun, and that it circles the Sun every 26 to 28 million years. The closest the companion star comes to Earth, they think, is from 20,000 to 30,000 Astronomical Units. (An Astronomical Unit—AU—is one Earth–Sun distance, roughly 93 million miles.) At that distance, the sun's companion is brushing against the very edges of the Solar System. But even the faintest brush by a small star would be a powerful shock in the peaceful regions of the comet cloud, certainly enough to pry loose a couple of hundred million comets and send them plunging in toward the center of the Solar System . . . and Earth.

"We are looking for it," Dr. Muller has said, referring to the elusive death star. But the real search, he added, "is in computer tapes" rather than in the sky. There are detailed catalogues of faint stars available for the perusal of scientists, and these catalogues are being fed into the computers for help in determining the likeliest candidates out there. None of the stars whose distance from Earth is known, is close enough to be Nemesis. So the computers must concentrate on those stars whose distance from us we *don't* know. Scientists have also told the computers to look for a slow-moving star, since most of the known stars which inhabit our corner of the Galaxy are fast movers.

Stormy, dangerous voyages through the interstellar seas as our sun voyages across the galactic plane: one theory. Nemesis, the death star: another theory. And now there is one more.

Planet X

The existence of a tenth planet has long been suspected by some astronomers. The gravitational effect of a tenth planet is one way of explaining certain slight deviations that have been observed in the orbits of Pluto and Neptune, the outermost planets. Planet X is the name given to the hypothetical planet by two professors from the University of Southwestern Louisiana, Daniel Whitmire and John Matese. According to the two researchers, the Oort cloud of comets, far beyond the orbit of Pluto, is spread out in a disk through the operation of gravity upon it. (Similarly, the known planets all circle the Sun in more or less the same plane, differing by only a few degrees.) It is within the disk of the Oort cloud, then, that Planet X circles the Sun.

Over the millennia, Planet X has swept a path for itself through the densely populated region of the Oort cloud. It is through this cleared, comet-free ring, or *annulus*, that Planet X travels as it orbits the Sun. So far, so good. With its own private road to travel, there is no need for Planet X to cause any trouble. But alas, space is not that simple, and the laws of trespass are not the laws of physics.

Planet X's orbit is inclined in such a way that once in a while—not often, but regularly—the planet brushes up against the edge of its path through the comets, scraping some of the comets hard enough to cause them to fall from their orbits and begin the long, ever-accelerating journey to the Sun. The Louisiana scientists believe that this regular scraping against the

149

comets occurs because Planet X's orbit is influenced by the gravity of other planets—just as its own gravity affects *them*. What this means is that the sharp curve at the end of the planet's elliptical orbit would be shifted slightly each time it rounded the Sun. This readjustment in its orbital path would mean that the planet would have to belt a few more comets out of the way from time to time.

How large is Planet X? The two astronomers guess its weight at anywhere from one to five times the mass of Earth. If its position is between 50 and 100 AUs from the Sun, which they suspect it is, then the planet would experience the orbital changes necessary to knock a few comets in toward Earth every 28 million years.

Can a theory like the theory of Planet X ever be proved? The answer is that it can, with observation. There is nothing like a picture of some new celestial face to convince a lot of skeptical astronomers that a new member has joined the family. But it will be difficult. The Louisiana pair is hoping to get time on the new Space Telescope—due to be launched by the shuttle in 1986—to find their elusive new planet.

CHAPTER 6

The Extraterrestrials

When the first bursts of data come hurrying back to Earth over the millions of miles of space, cometologists will not be the only people clustered around the computer waiting for the signals to be translated. There will also be a little knot, perhaps only two or three, who are interested in one of the more exotic adjuncts to the science of space. And that adjunct comprises the professors and researchers, the communications experts and the astrophysicists, the watchers and the wonderers of SETI—the Search for Extraterrestrial Intelligence.

These will not be lunatics, the SETI people. Isaac Asimov believes there is intelligent life out there, and that it is probably smarter than we are. And Carl Sagan has been involved in SETI for years. These are not sci-fi writers.

Dr. Robert Garrison is an astronomer at the University of Toronto's David Dunlap Observatory in Richmond Hill, Ontario. Dr. Garrison also teaches a course at the university called "Life on Other Worlds." This sounds like the kind of course that would attract half the undergraduate population, and include a lot of students with glassy expressions. But it is actually a rigorous, advanced course of study.

According to Dr. Garrison, astronomers interested in extraterrestrial intelligence will be watching the Halley's probes for certain basic information.

"We don't expect to find extraterrestrial creatures riding in on the thing," the astronomer says. "But we will be looking for evidence of certain complex molecules from interstellar space. The comets are the only deep-space probes we *have*. Comets are the only objects from interstellar space—from far enough out in the Solar System—that can tell us what molecules exist there.

"We are looking for clues to the origin of life."

Some of the comets originate in the farthest Oort cloud, which is, says Dr. Garrison, a third of the way from Earth to the nearest star. If there is evidence that complex molecules exist in interstellar space, molecules sufficiently complex to spark life, then SETI astronomers feel that the case for life on other planets is strengthened.

"We believe that our own Sun is a very typical star," Dr. Garrison explains. "We believe that there are many stars like the Sun, and that some of these will have planets. You don't *need* to have complex molecules from interstellar space to have life on those planets. But finding those molecules makes it easier to suppose you would find them on planets too."

And that is what Halley's may bring us. Just a little more evidence. Evidence of the neighbors.

They are out there, all right, and probably so much more advanced than we are that they've been watching us evolve for a million years. Where are they? They are elsewhere, or right here, going about their business until we have come far enough to possess the wit to comprehend what they have to say. As J. B. S. Haldane, the great geneticist, has said, "My suspicion is that the universe is not only queerer than we suppose, but queerer than we *can* suppose."

Are we all alone?

This wondering about life beyond our own solar system is not the exclusive preoccupation of sci-fi writers, hammering out life-and-death scenarios against the purple skies of Zondor. Some extremely canny members of our terrestrial species take the whole business as seriously as, well, business. When Stanley Kubrick was making *2001: A Space Odyssey,* he began to worry that intelligent extraterrestrial life would be discovered before his film was released. Kubrick's problem was that he had decided to represent extraterrestrial life in an abstract, surrealistic, nonspecific way, leaving it to your imagination to fill in whatever you chose. His worry was that before the film's release the real thing might arrive all by itself, with sixty fingers and half a dozen noses, and Kubrick's shiny slab of black marble would look about as mysterious as a blank tombstone. So apprehensive was Kubrick that he approached Lloyd's of London, the famous take-on-anything insurance brokers, looking for a policy that would insure him against the ruin of his film should extraterrestrial life show up before the date of release. You think *Kubrick* was crazy? Lloyd's wouldn't touch it!

Basics

Whatever they look like, whatever gods they worship or limbs they flex in the rainbow light of their strange worlds, whomever they love and hate, they are still made of the same stuff as man, for we are made of the stars. Our teeth and bones and skin, all have been created out of elements forged in the furnaces of massive stars. When stars are born, the vast prestellar nebulosity that spreads through space for millions of miles coalesces, gravitates, contracts, and finally presses in upon the center so heavily that the pressure and heat drive the atoms of hydrogen about in a wild melee. They crash and bounce and speed about until, at last, their velocity becomes so great that they begin to stick together. Four atoms of hydrogen stick together and form one atom of helium; but that one atom of helium weighs less than the combined weight of the four of hydrogen. What has happened? The rest of the mass, the part that has been lost, is released as energy, the energy of the thermonuclear reaction that is the star.

Later in its life—billions of years later—as the young star matures and turns slowly into an older member of the universe, it gets hotter, and so begins to produce atoms other than helium. The star begins to manufacture atoms of the heavier elements, elements that are the building blocks of planets like Earth. Carbon, silicon, oxygen, nitrogen, all handy materials for planet building, all come out of the blast furnace of the aging star. Infrared spectroscopy and radio astronomy reveal that some of the red giants are blowing off water, others graphite and silicates, all good usable stuff in the planet-assembly business, especially if you are trying to turn out a product that looks like Earth. One cloud of material being spewn into space by an exploding star, or *nova*, that the astronomers have been watching looks as if it's made of dolomite, the mixture of magnesium and carbon that makes up the Dolomite mountain range between Austria and northern Italy. The iron in your blood, the calcium that knits together your bones, the carbon on which are stamped the genetic instructions that make you look the way you do—all of this is billions of years old. All of it comes from the mighty stellar factories of space. You are made of the universe. Stars made you.

Visions

In *The Cosmic Connection* Carl Sagan notes that the idea that planets might harbor intelligent life was not common to our ancestors. Instead, they believed that the planets *were* intelligent beings: gods. Venus was the goddess of love; Mars ruled the affairs of war. Later, all this was to change, and even an astronomer and mathematician as great as Kepler could believe that every planet had its own population. William Herschel (1738–1822), a noted British astronomer, pushed this speculation further, proposing that the Sun itself was inhabited. Presumably these solar people would have been made entirely of gas, for if they were not before they took up residence, they would be soon after. These speculations seemed to peak around the time of the last appearance of Halley's comet over Earth, in 1910, when Professor Percival Lowell was stalking about the planet riveting audiences with his impassioned conceptions of life on Mars.

In the conclusion to his book *The Canals of Mars,* Lowell maintained that the habitation of Mars by beings of some sort was certain, although he couldn't say what they'd look like. Nevertheless, they were intelligent creatures. They had to be. Faced with the threat of their planet dying from lack of water, the Martians organized an elaborate and ingenious system of canals to carry the water remaining in the planet's polar regions to the areas that suffered drought.

The first thing that is forced on us in conclusion is the necessarily intelligent and non-bellicose character of the community which could thus act as a unit throughout its globe. War is a survival among us from savage times and affects now chiefly the boyish and unthinking element of the nation. The wisest realize that there are better ways of practicing heroism and other and more certain ends of ensuring the survival of the fittest. It is something people outgrow. But whether they consciously practice peace or not, nature in its evolution eventually practices it for them, and after enough of the inhabitants of a globe have killed each other off, the remainder must find it more advantageous to work together for the common good. Whether increasing common sense or increasing necessity was the spur that drove the Martians to this eminently sagacious state we cannot say, but it is certain that reached it they have, and equally certain that if they had not they must all die. When a planet has attained to the age of advancing decrepitude, and the remnant of its water supply resides simply in its polar caps, these can only be effectively tapped for the benefit of the inhabitants when arctic and equatorial peoples are at one. Difference of policy on the question of the all-important water supply means nothing short of death. Isolated communities cannot there [on Mars] be sufficient unto themselves; they must combine to solidarity or perish.

This is a wonderful vision of political harmony, but a spurious one. There are no canals on Mars, nor are there any Martians. There is no life as we know it. And yet this is not the vision of some jumped-up quack. Percival Lowell was professor of astronomy at the Massachusetts Institute of Technology; he was the director of the observatory at Flagstaff, Arizona; he was a fellow of the American Academy of Arts and

Sciences and a dozen other national organizations of stature. He was treated with respect by scientists around the world. He was, of course, also wrong. And every now and again there peeks through the writing of the old astronomer some little sign that he knew it himself.

The less the life there [on Mars] proves a counterpart of our earthly state of things, the more it fires fancy and piques inquiry as to what it be. We have all felt this impulse in our childhood as our ancestors did before us, when they conjured goblins and spirits from the vasty void, and if our energy continue, we never cease to feel its force through life.

He may have been wrong in the particulars, but much of what Lowell had to say sounds today like the speech of a man far ahead of his time. When he died in 1916, Lowell was just about the last astronomer to care a straw for planets at all. For the next four decades, astronomers gave themselves to the stars, as tremendous advances in the new field of astrophysics opened up stellar astronomy. So complete was the desertion of the planets and their putative creatures that just after World War II there was *only one astronomer in the whole of the United States* studying planets. Happily for those who like to think we are not all alone out here in the humbling vastness of creation, the search is on again.

Installing the Stellar Radio

Since 1950, developments in sophisticated instrumentation have enabled astronomers to probe the heavens with greater certainty, and what they have found there is nothing less than the stuff of life. More than that. Astronomers now believe that stars and planets go together, and that our own solar system is not just some fortuitous fluke of the cosmos, never to be duplicated anywhere in the "vasty void." Way out there, the smart money is now betting, some other star is wandering amiably through its galaxy, or maybe *our* galaxy, towing along in its wake its own set of planetary dependents like a string of bobbing kids. If these planets, just some of the trillions there must be, have developed advanced technological civilizations—and some of them must have—then they must have developed ways of talking to one another. They may even be sending out tentative signals to peoples as dim as ourselves, patiently waiting for us to grow up enough to answer back. What the scientists have done now for Earth is installed the radio. Now all we have to do is fiddle with the dial until we find the right station. And then we can listen to the clever, coded signals broadcast across the plunging valleys of space to where we listen at our clumsy sets.

At Arecibo, Puerto Rico, Cornell University runs the facility that controls the mighty 1,000-foot-diameter radio telescope owned by the National Astronomy and Ionosphere Center. Three tall stanchions stand atop sharp, steep-sided hills that surround a broad depression in the mountainous jungle of the island. Slung from these stanchions on strong cables is the focusing mechanism of the telescope. Beneath it, the great dish sits upon the ground, laid into the carefully graded depression. It is a breathtaking structure, suggesting that man at last is ready to listen to the universe, and that it is only a matter of patience and time before we hear what it has to say.

The phone lines are in. If they're ready for us, we can talk to them, all right. We can pump out a message that will carry its clear little signal of intelligence tens of thousands of light-years away. The problem is finding the right number to dial.

Defining the Mission

In 1971, a group of skeptical but inquiring men met at Byurakan, in Soviet Armenia. The men were astronomers and physicists, historians and anthropologists, cryptographers and computer experts. Their meeting had two patrons, joint sponsors. One was the National Academy of Sciences of the United States, the other the Soviet Union's Soviet Academy of Sciences. In a world so often poisoned by enmity between governments, this was a cheering event. The scientists gathered in Byurakan decided that the odds on there being intelligent creatures in the cosmos who could respond to us, and our own technological ability to establish contact with such creatures, were both high enough to call for an attempt. Here are a few of the conclusions the scientists reached:

- Recent scientific advances make it possible for man to go beyond the state of merely dreaming about extraterrestrial counterparts in the universe. We are now in a position to explore. We are able, in short, to move from theory to experiment.
- The consequences for mankind of establishing contact with a civilization beyond our world are so charged with possibilities that we have a duty to attempt the feat. Who knows what we may learn of philosophy, of medicine, of some new field of thought that could revolutionize the way we run our own beautiful, but flawed, planet.
- There ought to be inter-nation cooperation so

that the search for other intelligence is a search
by the whole globe.
- Keep the projects modest in scale. Money is
tight.
- Maintain the search for other solar systems.
- Just keep looking . . . for anything.

Where to Meet E.T.

In *The Cosmic Connection* Sagan likens the search
for extraterrestrial life to two strangers who have never
met agreeing to meet in New York City, but arranging
no time or place for their rendezvous. At first the task
seems daunting, even impossible. But is it? You would
not stand, say, at the corner of Lexington and Seventy-
ninth. You would look instead in places well known to
everyone, like Rockefeller Center or Times Square or
the Empire State Building. The possible places, Sagan
maintains, would be more in the dozens than in the
millions. You would know which places were the most
important, or likely, places for an assignation. All that
would remain would be to patrol those spots and
eventually you could make contact.

That analogy, says Sagan, works in space. They know
we know that hydrogen, the most common atom in
space, emits at a frequency of 1,420 megahertz. In
other words, there are basic frequencies that any
extraterrestrial intelligence capable of communicating
with us by radio astronomy would recognize at once.
We are not going to receive a signal saying, "Hey! Do
you eat Znongburgers?" The creature at the other end
of the line will more likely be tapping out "helium" or
"water" in the universal code he knows we know he
understands.

What, Sagan asks, are the likely Times Squares of
space? We could begin with stars as old as or older than

our own. We might even aim our scans at the regions of the black holes. Some of these might turn out not to be black holes at all, but staggering projects of astroengineering erected by some great and powerful interstellar civilization, masters of forces strong enough to bend the curves of time and space. Another approach might be to bypass the stars of our own galaxy, the Milky Way. There are about 200 billion stars in our galaxy to sort through, and we could strike out a couple of million times without raising so much as a peep from possible residents.

Instead of this laborious process of aiming at one star, then another, we could point our radio telescopes beyond, perhaps to the great galaxy in Andromeda, M31. M31 has about 200 billion stars as well, give or take a billion, but they are so far away that from here they appear all bunched together in a manageably sized blob of light, and we could cover every star in M31 at once. That way, if only a couple of those 200 billion stars are warming planets with intelligent life, we ought to hear whatever they are saying.

Fearing the Enemy

One of the abiding fears of those who would be more circumspect in the search for extraterrestrials is that they might turn out to be just as bellicose and unprincipled a pack of scoundrels as we are. Oh sure, they say, there's that faintly soppy vision of some balloon-headed creature whose yard-wide noggin is crammed with galactic compassion and a kind of patronizing superintelligence. This conception envisions a Saint Francis of Assisi with a benevolent smile and one of those little wandlike zappers that calm the raging hearts of the beasts of Earth. But what if instead, the cynics continue, our pathetic attempt to shake the hand (if hand there is) of our extraterrestrial brethren goes awry?

The most likely vehicle of communication will probably be nothing more menacing than a radio wave. Whatever it carries, it will be a message from the past, for the civilization that sends it will have had to possess the technology to transmit long ago. If it comes from a civilization 100 light-years away, the message will already be 100 years old. If from 1,000 light-years, the signal will arrive at our planet across the gulf of one millennium. The civilization that sent it might have already perished, sending out one last dying warning to worlds it had never met. Whatever the message, it will come from a civilization more advanced than ours, since they will have sent it before we ourselves possessed even the technology to receive it. One hopes that such a message would be framed in such a way that it would be easily understood by beings without the language of the sender. If there are such communications knifing through space toward us, they probably come with instructions to help simple-minded Earthlings understand them. There may even be a lively galactic tradition of passing on friendly tips to newly evolved intelligent beings, like us. Even if we haven't yet managed to snare a signal ourselves from the masters of Andromeda, perhaps we should be keeping a close eye on some likely spots in our own solar system, ready to beam out an encouraging word to them when they're ready to listen for it. Take Europa.

Precious Life

The moon Europa is a mysteriously smooth and ice-covered satellite of the giant planet Jupiter. NASA scientists now believe that Europa's ice is thin, somehow melted, or thinned, by her own heat. The surface of Europa is marked by dark bands on top of the ice. The NASA scientists think these bands are laid down on the surface when water from huge geysers gushes up from below and deposits some marine material on the frozen seas of the circling moon. When the ice breaks as the geyser drives through the surface, enough sunlight enters the depths beneath to allow plants to photosynthesize. Beneath the thick ice of our own Antarctica mats of algae survive, similar to those envisioned.

Who knows what life may evolve on Europa? Perhaps we should send them a satellite, a package of information to circle the little moon for the next few million years, waiting for some creatures to evolve the intelligence to pull it down and read the contents. Perhaps, dreadful to contemplate, we might tell them about our factional Earth, maybe by then destroyed in a battle that leaves us a desolate lump of rock. Staring back at the wasted third planet from the Sun, and reading our descriptions of this green and lovely home, the Europans might be moved to resolve any differences of their own by flipping coins. Let's include a few.

SETI

Meanwhile, we must listen for our own messages. Here is what the National Academy of Sciences has to say:

> Astronomy has shown that there are enormous numbers of stars like the Sun and that the abundance of chemical elements is much the same everywhere. It seems possible, therefore, that there are habitats for life scattered throughout the universe.

Now at last, SETI—the Search for Extraterrestrial Intelligence—is under way. It is modest, but at least it's a beginning. NASA received $1.5 million in 1983, the first year of a planned five-year program to develop the right equipment to process whatever signals we receive. Each year of the program, up to 1987, SETI will receive another allotment to help it on its way. NASA will be able to stretch this money by using some of the radio-telescopic facilities already in existence for talking to the probes in deep space.

Initially, a multichannel spectrum analyzer will scan 74,000 frequency ranges simultaneously. This is just for starters. Once the bugs are ironed out of the humble 74,000-frequency-range model, SETI plans to replace it with one that will simultaneously scan 8 million channels. This facility will be located at the deep space monitoring network's station at Goldstone, California. Stanford University, which is developing the searching technology, is aiming at a device that will *listen in on 100 million frequencies at the same time*. This scale of eavesdropping only makes sense when you remember that there are thousands of billions of stars that might maintain solar systems with life-supporting planets.

Whatever is gathered by the Stanford team will be fed into a computer at the Jet Propulsion Laboratory (JPL) of the California Institute of Technology. JPL is located in Pasadena but operates the facility at Goldstone. This computer is programmed to ignore the tremendous amount of blathering poured into the atmosphere by man. The scanner itself is constructed to pick up only radio emissions of a very narrow bandwidth. These bands would, in fact, be narrower than any produced by the natural emissions of space. Or so we hope. What we might be able to pick up in this way, for example, is the TV traffic of another planet.

When a TV station puts out its signal, it transmits a narrow-band frequency carrier in a horizontal direction. But the signal doesn't follow the curve of Earth. It clatters into your home by whatever means you collect it, and then simply keeps on moving in a straight line, out to the horizon and beyond, straight into space. When many of these signals are grouped together—say, from all the TV stations in the New York metropolitan area—the effect is to beam into space a combined narrow-band emission of considerable strength, which sweeps out across space in a curving line as our planet revolves. If the same domestic traffic is spilling into space from another planet and we are tuned to the right narrow band, then we might easily pick up a celestial *Cosmos,* featuring a greenish, pointy-eared version of Carl Sagan, wondering to his viewers whether *we* exist.

While the project run by SETI is ambitious enough to aim at sweeping the entire sky, it will also concentrate on 773 stars within eighty light-years of Earth. These stars are close enough to the Sun in size and age to make it possible that they have solar systems of their own, and planets where life may have evolved.

Meeting at the Water Hole

They are searching the heavens right now, and this current program of SETI will cover ten times as many stars as all thirty-six similar projects since 1960. It will monitor a range of frequencies 3 million times greater than previous searches, and will peer at stars in the Milky Way as well as some of the nearer galaxies. But they are not just searching blindly, those scientists out in the desert and hovering over the computers in Pasadena. The scientists in charge of the deep-space network of radio dishes involved in SETI are concentrating their search near what they call the "water hole." This is the part of the radio spectrum that includes the frequencies emitted by hydrogen and hydroxyl, which produce water when they are combined. Other civilizations, so the reasoning goes, might find the water hole a logical place to meet us. In other words, the water hole might be one of Sagan's Times Squares of space.

Of course there is a simpler way to accomplish all this, which does not seem to have suggested itself to Carl Sagan or any of the other fellows at SETI. We just sit around and wait for someone to show up and take us for a spin.

250 Million Civilizations

In his book *Extraterrestrial Civilizations*, Isaac Asimov maintains that the very size of the observable universe, 1 billion trillion stars—1,000,000,000,000,-000,000,000—makes it close to certain that there is intelligent life riding some planet around one of them. Be pessimistic about it if you wish, says Asimov, and put the odds at one in a billion that a technological civilization is out there somewhere. Fine; at one-in-a-billion odds, there are 1 trillion technological civilizations in our universe. Blazing his logical trail through these fertile possibilities, Asimov leads us to his own conclusion with masterful reasoning.

Here's how it goes.

The number of stars in our galaxy, the Milky Way, Asimov puts at 300 billion. Elsewhere in this book you may read other figures for the Milky Way, such as 200 billion. This is a discrepancy of 100 billion (100,000,000,000). Don't worry about it. It's not money, only stars, and the accounting employed by astronomers is gloriously loose.

Of these 300 billion stars, Asimov says, 280 billion will be of the slowly rotating variety, and thus likely to possess planetary systems of their own. Of course, not all of these stars will be suitable for creating a climate favorable to life, but something over 25 percent of them are close enough to our Sun in size and composition to act as life generators and supporters. So let's say 75 billion of the planetary systems in the Milky Way are orbiting suns more or less like ours—friendly.

To narrow the field further, not every one of these sunlike stars will be blazing forth its heat into a great enough volume of space to benefit its circling planets. But a little over two-thirds of them ought to be, or 52 billion.

Now, it takes a special kind of sun first to create and then to nurture life as we define it. This sun must have certain rocks and metals. Therefore it will be a star in the outer regions of the galaxies, like our own. Only about one-tenth of the last figure, or about 5.2 billion stars, are suitable. Of these, only half, or 2.6 billion stars, will have planets located close enough to benefit from their effects. Of those, only half will have an Earth-like planet in an ideal orbit, and half again of those plants will be habitable.

We have narrowed the field from 300 billion stars to 650 million planets. Asimov is not through yet.

Only 92 percent of the 650 million planets where life may exist are old enough for life to have developed. Let's say, then, that there are 600 million planets in the Milky Way where life exists. But not all of these planets are old enough to have developed multicellular life. A little over two-thirds are. Thus, 433 million planets in our own galaxy have developed far enough to possess multicellular life.

Hang on; we're closing in.

Not all of those 433 million planets in the Milky Way with multicellular life will have had that life going long enough to produce as rich a zoosphere as Earth's. Asimov estimates that 416 million of them have produced such a diversity of life. And of those, 60 percent are old enough to have developed a civilization. We have arrived.

There are 250 million planets in the Milky Way with civilizations.

It is all so seductive—and possible! Isaac Asimov did his fanciful calculations just for the Milky Way. But the Milky Way is only one galaxy. Beyond it there are clusters of galaxies, and beyond those huge assemblies of clusters, superclusters. Until recently it was believed that galaxies were scattered randomly through space. One analysis now suggests that at least two of these unimaginable gatherings of trillions upon trillions of stars are linked. The Lynx-Ursa Supercluster and the Perseus Supercluster are one single highway of galaxies, leading no one knows where.

And so we await the first bursts of tantalizing data from the bravely voyaging probes. For if Halley's bears a tale of life—even of *possible* life—then what might those galaxies hold!

CHAPTER 7

Epidemics from Space

It always happens first to that stenographer in the next office, the young woman who eats nothing but fruit and nuts, who breakfasts on a loathsome mess of bran and wheat germ, and whose skin, in consequence, glows like the morning sun. One day, she is simply not there, and you learn that she's caught some "bug." A few days later you could fire a cannon down the rows of desks and scarcely graze a soul. This is the handiwork of that energetic little predator, the influenza virus. It thins out whole classrooms, empties factories, drives the populace to bed, and sometimes, very ruthlessly, it kills people. Today, two serious men contend that this murderous bug descends from space. They are Sir Fred Hoyle and Professor Chandra Wickramasinghe.

Sir Fred Hoyle has long been famous throughout the world for his many and daring contributions to the intriguing squabbles that constitute professional astronomy. In 1957 he was made a fellow of the Royal Society. He became an honorary member of the American Academy of Arts and Sciences in 1964, and in 1969 a foreign associate of the U.S. National Academy of Sciences. He was knighted by Queen Elizabeth II in 1972. Chandra Wickramasinghe is recognized as an international authority on interstellar matter. His first collaboration with Sir Fred was in 1962, when they proposed a radical new theory on the origins of interstellar dust. He is professor and head of the department of Applied Mathematics and Astronomy, University College, Cardiff, Wales.

These two distinguished scientists believe that all of the viruses and bacteria that cause and spread infectious diseases in plant and man alike were delivered to our terrestrial doorstep fresh from the germ factories of space. This is what they attempt to prove in their fascinating book *Diseases from Space*, and they make the attempt with plenty of argument.

Learning to Kill

Right off the top, Sir Fred confronts one of the more lethal shoals on which his thesis might founder. Whereas bacteria, an undiscriminating class of germ, just mosey into whatever system they intend to attack and begin to munch away, a virus is more selective. It attacks a specific cell, and so must locate that specific prey if it is to survive. The objection to Hoyle and Wickramasinghe goes like this: How does a virus, tumbling about out there in space, acquire the specific weaponry it needs to engage in battle and finally defeat the Earth-bound cell that is its quarry? The answer offered is a chilling one.

Bombarded by these space invaders for hundreds of thousands of years, the organisms of Earth, such as man, have *learned to receive them*. Why would our bodies tolerate such an adaptation to illness? Because, although a given disease might be bad for one man, or several men, it might turn out to be a good thing for mankind as a whole—because it would weed out weak specimens.

Further, Hoyle and Wickramasinghe argue in favor of biochemical unity throughout the Solar System. What this means is that there need not be different forms of the basic structures of life, but that all might match, or closely resemble, our own. They reason it this way: All biological multiplication is explosively rapid. Take the term of the common cold. Between the time you catch the cold and the moment of the first trumpeted sneeze, the little virus that started the whole thing has multiplied *ten thousand million times*.

Take another example: The pomace fly, Drosophila, can produce young twenty-five times a year, 100 offspring each time. If there were no predators or other natural killers, and each successive generation could continue to produce offspring, in one year there would be a ball of pomace flies *96 million miles in diameter, the flies packed a thousand to the inch*.

What Sir Fred concludes from all this natural fecundity is that whichever system first managed to evolve some kind of biological life reproduced so fast that it beat any other experiments to the punch, seeding the whole Solar System with its version of life and overwhelming the competitors. Thus *any* germ from space would be compatible with Earth. In other words, it would always find something good to eat.

The Ride to Earth

The life that peppered Earth in the form of molecules, and those basic biological grandfathers, the atoms, were all gathered into the newly forming Solar System as it swept out of some great cloud of interstellar gas five thousand million years ago. Within the infant Solar System, the seminal stuff of the planets was already settling into orbits around the boiling Sun. Dragged along at the farthest edge of the Sun's influence were the comets, pacing a great majestic dance in numbers so profuse they would have looked like one vast, unbroken cloud. Hundreds of billions of these early, cometlike bodies swarmed within the giant cloud. They were frozen cores of simple substances, such as water. On top of them, the theory runs, lay a mantle of organic material perhaps half a mile thick. As they milled in space, some of these bodies collided, smacking into one another like dancers on a crowded floor. These collisions sparked reactions which, eventually, caused heating and melting deep within pockets protected beneath the cometary surface. Complex organisms, perhaps polysaccharides for energy, dwelt and bubbled in the black, encircled caves, rolling along the celestial path of time and space that carried them finally to Earth, sowers of ourselves and all we see.

Something like ten comets a year come bowling into our terrestrial vicinity, and a few of them will bear these pockets of living or ready-to-live germs. As the comet tears into the last lap and approaches the Sun, it begins to shed part of its mass in the ragged current of the solar wind, and as the last layers peel away, the germs spill out into space. Some of them will be destroyed in the general savagery of unprotected space, but some will make it into our atmosphere, where their soft landing is assured. If their spacefall takes them to the Moon, on the other hand, they will not be so lucky. With no fat atmosphere to cushion their fall, they will simply spatter onto the lunar surface so hard they will turn to gas. But on Earth they have nothing to worry about.

This fall of pathogens—disease carriers—is irregular. They are blown about by weather, and affected by the

Earth's rotation and the path of whatever comet is bringing them in. So if, let's say, Halley's is fetching along some nasty surprise in the way of a new influenza virus, it's just possible that the whole batch will be dumped onto Albania, and we will hear no more of it, ever again. Conversely, the pathogens might be spread by the wind in a pattern that will cover several continents. Still another possibility is that particles will remain aloft for years, sneaking aground when we simple Earthlings have long forgotten about the comet that brought them. We already know from the study of volcanoes that fine dust can remain in the stratosphere for a decade. The unevenness of pathogenic dispersal over the face of the earth accounts, in Sir Fred's opinion, for stories from the days of sea travel of voyagers being stricken by epidemics days after leaving port. They would simply have sailed into a patch of some plaguey junk that had not blown ashore.

Random Strikes

If it is true that germs fall from space and are spread in patches about the globe, then it ought to be true that people who travel run a greater risk of encountering a patch of infection than those who stay home. And if this is true for man, it ought to be true for animals. Thus birds, the greatest travelers of us all, ought to have evolved more efficient immunities than we more sluggardly creatures. In fact, this is true, at least as far as flu goes. Birds do have better immunity than man.

So what about the efforts of such august disease fighters as the World Health Organization, which claims to have wiped out smallpox once and for all? Nonsense, says Sir Fred. If the comet that dumped the stuff on us in the first place left a lot, then it's here until it runs out all by itself. And, of course, it could be replenished before that ever happens. One current affliction, the virus herpes simplex, has a pool of support that scientists estimate will last thousands of years. This is not to say that medicine won't find some way to rid us of the effects of herpes, only that any such treatment won't mean that herpes itself has cleared out for good.

To support the contention that diseases come and go on Earth at the bidding of celestial phenomena beyond man's control, Hoyle and Wickramasinghe quote a long, detailed, and gruesome account culled from Thucydides' description of a terrible plague that struck Athens in 430 B.C. One of the symptoms was a fever so intense that the suffering victims, desperate for any surcease, threw themselves into the cisterns. Other symptoms were the loss of fingers and toes . . . and even of memory! Maintaining no reason to doubt the accuracy of the reporter, the two authors conclude that since the symptoms do not fit any disease which we know today, the awful plague represented a single, vicious foray by some space germ that has never reappeared.

Tracking the Virus

Is it possible to establish the truth of these chilling suppositions? Thousands of tons of micrometeorites fall upon the surface of our planet every year. To sift through all this debris as it floats to Earth, winnowing out the plain dirt in search of the lurking germ, would be a very daunting task, indeed.

What about ice cores? Frozen into the polar ice caps for centuries must be a perfect record of the germs that fell to Earth over all time. All the diligent searcher would have to do is drill himself a core, draw it out, and count what he finds. But here again, there are no techniques available for quickly picking out a virus from a surrounding ton of ice. And there would be no way to determine for certain whether any particles located in the core necessarily came from space. They could as easily have blown there from some other region of Earth. No, the best way to find a virus is to find a sick person.

Now, from the time a few particles of virus—let's say of the common cold, which is Sir Fred's example—wander into your mouth and begin the work of multiplying, until they have become thick enough for the sneezing to begin, a few days must elapse. In that short time, the virus must increase to a density of 10 billion particles per cubic centimeter. Even as energetic a germ as the cold virus takes time to make its laborious way in this fashion: infecting, incubating, building up to invasion strength, and finally hitting the beaches. And yet there are plenty of examples of the spontaneous outbreak of the common cold among many people at the same time. This, our theorists maintain, is because they all walked together into the same patch of falling cold germs, and so caught the virus at the same time, *not from one another.* Why is the cold so universal and so unceasing in its attacks on man? Because as Earth rolls on through space, picking up its daily tons of dirt, some of the hundred or so different kinds of cold germs are stirred into the batch. When cold germs marched on Earth, they came in strength.

Hospital of Ghouls

To buttress their theory, Sir Fred and Professor Wickramasinghe cite a study into causes of the common cold carried out by the British Medical Council. The

Common Cold Research Unit set up its ghoulish camp at an old U.S. Army hospital near Salisbury. Then they set about enjoying themselves, or at least the sadists in charge did. After taking hot baths, participants in the study were made to stand around in cold, drafty corridors, wearing only bathing suits. And they had to stay in these halls, shivering. until they could bear it no longer. Only then were they allowed to dress. But even dressed, they had to keep their feet in wet socks for another couple of hours. By now, even the British would feel cold. And yet not one of those who simply reduced himself to shivering misery actually caught a cold. It took more. *Only those who were finally slipped a shot of cold virus caught a cold.* And yet people do catch cold in the winter, and more often than at other times of year. You thought this was because people got colder in the winter, and you were not alone. But not so, says the theory. There are more colds in the winter because winter storms, usually violent, shake up the atmosphere. When they do, out fall the germs.

Further in pursuit of knowledge, the Common Cold Research Unit embarked on other experiments, this time concluding that colds are rarely, if ever, spread by contact—good news for Sir Fred.

Lunatic Island

Not content with the old army hospital near Salisbury, the scientists decided to maroon their victims. They located a small, isolated island, Eilean nan Ron, lying a mile and a half off the Scottish coast.

Soon "invaders," their noses running with the vile stamp of their disease, were creeping about the island attempting to infect their coldless fellows by an onslaught of the most determined chumminess. It didn't work. Breathe in their faces though they would, the cold sufferers could not afflict the others with the virus. There was one case of cold reported ashore, but Sir Fred attributes it to a spacefall in the area, perhaps triggered by a spectacular display of aurora borealis recorded at the time by the strange denizens of Eilean nan Ron.

Ancient Enemy

Outbreaks of what was probably influenza have been recorded for almost a thousand years. The "English Sweats," which struck five times between 1485 and 1552, may have been a deadly form of the disease. The toll was in the millions. Detailing the history of the disease, Sir Fred records one early description contained in a letter from Lord Randolph, in Edinburgh, to Lord Cecil. The year is 1562.

May it please your Honor, immediately upon the Quene's [Mary, Queen of Scots] arivall here, she fell acquaynted with a new disease that is common in this town, called here the new acquayntance, which passed also through her whole courte neither sparinge lords, ladies nor damoysells not so much as either Frenche or English. It ys a plague in their heades that have yt, and a sorenes in their stomackes, with a great coughe, that remayneth with some longer, with others shorter time, as yet findeth apt bodies for the nature of the disease. The queene kept her bed six days. There was no appearance of danger, nor manie that die of the disease, excepte some olde folkes. My lord of Murraye is now presently in it, the lord of Lidlington hathe had it, and I am ashamed to say that I

have byne free of it, seinge it seketh acquayntance
at all men's handes.

The catalogue of human misery at the hands of
influenza has many entries. In 1580 Asia, Africa,
Europe, and the Americas all suffered . . . the whole
planet. In modern times, the pandemic of 1889–90 was
the very figure of death. It began in Bukhara, Russia,
in the early summer of 1889. By autumn it had spread
from one end of that vast empire to the other. By
winter, the populations of Western Europe and North
America were dying in the hundreds of thousands.

In the ravages of the epidemic of 1917–19, whole
villages in India were simply erased, all the inhabitants
dead at a stroke. And yet on St. Helena, in the South
Atlantic smack in the middle of the shipping lanes
leading around the Cape of Good Hope, not a man
perished. This is astonishing if you believe in human
contagion, for certainly ships on their way home from
the steaming, desperate, stricken ports of the Indian
subcontinent would have put in at St. Helena by the
score.

Dread Caprice

In the United States, the apparent arbitrariness with
which the killer strikes is registered clearly in a compar-
ison of the death rates of Toledo and Pittsburgh.
Normally, these two cities experienced mortality rates
within a few percent of each other. But when the
disease struck in late 1918, it was Pittsburgh that took
the full force of the blow. Its death rate for that part of
the year was 400 percent higher than Toledo's. If the flu
is spread by contact among the human population,
surely no such great a discrepancy could ever occur in a
country as mobile as the United States.

As further evidence of the capricious behavior of flu, there are reports suggesting a substantial infection of baboons in South Africa during the pandemic of 1917– 19. It is difficult to imagine how these aggressive and unfriendly creatures could ever have contracted the disease from man, with whom they are not exactly on the most cordial of terms. Similarly, both Canada and the United States reported extraordinary numbers of deaths among moose, elk, deer, and antelope during that outbreak, apparently simultaneous with the outbreak of the disease among humans.

Influenza, it seems, does not stalk its human victims by trailing them along the routes of international travel, stealing onto aircraft or stowing away aboard ship. When the so-called Hong Kong flu of 1968 stabbed ashore onto the continental United States, it hit, naturally enough, in California. But it did not appear first in Los Angeles, or even San Francisco, the places where one would expect it if it had been following along from victim to victim. No, it contrived to land in Needles, a desert town in the interior of the state, far from any port, far from any international airport, far from most people.

It took weeks before it burst loose.

In one short and terrifying week it leapt the whole continent and appeared in New York City. By the next week, it had appeared in Chicago. In one swift and sudden move, it plunged down the country to seize Texas. From there, ignoring a Miami thronged with travelers at the peak of the Christmas season, the marauder jabbed north again to the Great Lakes. This was no disease dragging itself around the nation on the coattails of the citizens, conclude Hoyle and Wickramasinghe, but a pathogenic enemy from space, depending for its travel on the vagaries of the weather. The recurrence of some forms of flu, which others ascribe to the survival of the virus in animal hosts, Sir

179

Fred puts down to the persistence of some of the pathogens in space, and Earth's plowing through them a second time, or a third, whenever our orbit carries us through the viral clouds.

Striking the Schools

Casting about for some illustrative creature more absorbing than the South African baboon, the two astronomers turned their speculative gaze closer to home. There, it settled upon that intriguing species, the British schoolboy.

St. Donat's is an international preparatory school, built around the ruins of an old castle on the Welsh coast. The boys' sleeping arrangements consist of eighty-five dormitories, each holding four beds. Here's how the disease hit St. Donat's: thirty-five dormitories, one case each; five dorms, two each; one dorm, three sick. The remaining forty-four dormitories escaped the flu entirely!

This was the point at which Sir Fred and Chandra Wickramasinghe regarded their case as made. Here it was, the strongest proof to date that disease, at least this one, was not transmitted by human meeting human. If it had been, any dormitory where one case appeared ought soon to contain four cases. And any whole school where the virus squeezed in should in a matter of days be staring out at the world with uniformly forlorn faces. Yet more than half the dormitories went untouched.

At this point, lesser men would have shuffled their data together, crammed the package into the mail, shucked off their lab coats, and headed for the pub. But Hoyle and Wickramasinghe wanted more victims. They found them among that progeny of the elite, the boys of Eton.

Eton College, near the royal castle of Windsor, has furnished generations of leaders to a nation now no longer quite as keen on having aristocrats at the helm. There were 1,248 pupils at Eton, living in twenty-five "houses," or residences, when Sir Fred began to sift through the records, and 441 of them had come down with the flu. To demonstrate how wildly uneven is the attack of the disease, the researchers established a standard. If there was any orderly transmission of the disease at all, then it ought to be possible to calculate how many cases would show up in each of the houses of the school. One of the twenty-five houses, College House, ought to have had twenty-five ailing lads abed with the groans. It had only *one* case! The two astronomers figured that the odds of this being so were one in a billion. What's more, when they computed in all the other wild fluctuations between houses and the percentage of students afflicted, they concluded there would have to be another 100 quadrillion (100,000,000,000,-000,000) epidemics before the same results appeared! This is not the behavior of a predictable disease. But it gets more capricious still.

Eton suffered an influenza attack rate of 35 percent. A couple of hundred yards down the street from Eton is another school, St. George's. At St. George's, *there was not one single case of flu.*

Byzantium to America

In A.D. 540, during the reign of the emperor Justinian, plague swept the Middle East, North Africa, and Europe. One historian has reckoned the dead at 100 million. The infamous Black Death of the fourteenth century struck throughout the world, again taking millions of victims. The most recent heavy blows from this monster of death were dealt in our century, with an

181

outbreak that ended in 1917 claiming 13 million people in India.

There are all sorts of bizarre and loathsome tales associated with the bubonic plague. One has the Tartars, never a fastidious people, using plague victims as ammunition. Besieging an Italian base in the Crimea in 1347, the story runs, the Tartars loaded their catapults with plague corpses and lobbed them into the midst of the defenders. This spread the disease to the Italians, who promptly took it home by ship. But there's a snag to this neat account. We know that the fleas that carry plague leave the dead bodies as fast as they can in search of warmer homes. It was the wicked Tartars, not Italians, who would have supplied fresh housing for the deadly pests.

It is possible to trace the spread of one epidemic of plague in the Middle Ages as it moved north from the Mediterranean. There seem to be progressively more northerly outbreaks at intervals of six months. One school of thought attributes this to the steady progress of the plague-bearing rats north, a progress slow but relentless. The two astronomers, Hoyle and Wickramasinghe, shake their heads in scorn, maintaining that the rats, sick to begin with, would never have made it past the Alps. No, it was not on the feet of ailing rodents that the scourge ascended Europe as if it were climbing a ladder of death. It simply fell here and there, as arbitrary as a demented murderer.

One last example to close the case, this one right at the front door. In 1976 in Philadelphia, a nameless, baffling killer rocked the best minds of medicine, who vainly tried to track it down. Dubbed Legionnaires' disease, it laid low 183 victims, slaying 29 of them. But there were 3,683 delegates to the convention of the American Legion where the disease made its first appearance on Earth. All of these men were in constant and close contact with one another for the period of the convention, thronging the meeting rooms and the crowded convention floor where the main events took place. If the bacillus, the ugly, mysterious, lethal, and utterly strange bacillus, were spread by contact, how could a scant 183 men have fallen to it? The answer, one answer, is that all of those men had one thing in common. *Each of them had been standing outside the hotel watching the same parade.*

One small cloud of germs: a new disease from space?

The Germ Debate

Many of the scientists who have plunged into the debate about organisms from space—whether on the side of the believers or of the skeptics—are eagerly awaiting the arrival of Halley's to furnish evidence to decide the issue. It is hoped that the probes planned to intercept the comet in 1986 will discover whether there is any life-nurturing material aboard. But in the meantime, it was too much for even people as phlegmatic as scientists to endure the long wait without at least one more free-for-all in which to record their stances. And so on November 11, 1983, a scientific symposium entitled "Are Interstellar Grains Bacterial?" met in London.

For the first time the opponents could meet face to face. Organized by the Royal Astronomical Society, the meeting herded proponents and detractors, astronomers and biologists, into the same lecture room and let them go at it. Chandra Wickramasinghe opened the discussion with the presentation of a paper in which he outlined his case that interstellar grains are biological organisms, and not simply particles of ice and minerals. This is the basic premise upon which Wickramasinghe and Hoyle's work is built.

First of all, Wickramasinghe noted that bacteria fit

perfectly into the right-size scale to explain the extinction of starlight. In other words, if the clouds of interstellar matter that diffuse and obliterate the light from some stars are fashioned from microscopic grains, as they are, then these grains are the same size as terrestrial bacteria.

Next, Wickramasinghe moved into a discussion of spectroscopic evidence. Spectroscopy is the study of spectra, the photographic "pictures" painted by the emission of energy from a radiant source, arranged in order of wavelengths. Translated, this means looking at pictures of bars of color. Each substance in creation produces a different picture. If you hold a prism up to the window, it will produce a spectrum, the beautiful colors of broken sunlight arranged in bars. Similarly, with elaborate equipment, scientists can examine the spectral profiles of any substance that radiates energy.

Wickramasinghe noted parallels between certain characteristics of a spectrum available from space and the laboratory data for *E. coli,* one of the most common bacteria on Earth. There was a good deal of heat as the scientists batted this one around. One of the participants, a chemist from Sussex University, agreed that the similarities between *E. coli* and the spectral profile from space were remarkable, but not unique. And the chemist H. W. Kroto insisted that no self-respecting laboratory spectroscopist would ever try to make his case on the identification of a single characteristic. The similarity that Wickramasinghe had noted, Kroto said, might also match other organic molecules.

Supporting the germs-from-space side, D. J. Kushner of the University of Ottawa noted that although the evolution of biological organisms under the battering extremes of low temperatures and low pressures that prevail in space was not very likely, those organisms might be able to survive if they had existed *before* they encountered the deep freeze of space. This fits the Hoyle-Wickramasinghe model. The comets received their mix of organic material far back in the seething chaos of the proto–Solar System. In the violent domain of those early millennia, some of that material was driven into womb-like, protected pockets deep within the comets, safely stowed for any journey. The evolution was in the boiling soup of our nascent Solar System, or in the protected world of the comets' interiors. All the bacteria must do is survive the fall to Earth when they are finally blasted free as the comet approaches the Sun.

But there is other evidence to support the theorists, and it is close at hand. Earth receives a more or less steady supply of cometary material, in the form of meteoritic debris. Certain structures found in samples of meteorite resemble fossil bacteria. This ties in to one of the abiding objections to the purely evolutionary model of the origin of life on Earth. And that is that fossil bacteria have been identified in Earth-bound rocks as much as 4 billion years old. Four billion years old is getting mighty close to the age of Earth itself, and does not really leave a lot of time for the fossilized bacteria to have evolved. Did they arrive on a comet? In the turbulence of those formative years, they easily could have. And if they arrived then, why not now?

Excitement mounts in the scientific world as the famous visitor draws closer and man prepares to take his first really close look at the hairy star. Already there are signs that the comet may possess the capability to support life. A preliminary look at the spectroscopic data from Halley's has indicated that the comet's icy surface is very red. This discovery was made by astronomers at the Kitt Peak National Observatory in Arizona, using the observatory's 158-inch telescope. In April 1984, when the observations were made, the comet was halfway between the orbits of Jupiter and Saturn.

At that distance, the comet was not close enough to

distinguish the spectral signatures that would tell astronomers what elements or molecules the comet comprised. But the red color did suggest that the comet's frozen nucleus was wrapped in material rich in complex organic molecules.

Something up there might be alive.

CHAPTER 8

Waiting for the Comet

It is a very different world that greets Halley's today than the veteran traveler met seventy-five years ago when it last blazed into our skies. For one thing, Earth is a much brighter place, and this presents special problems. When Halley's swung past in 1910, America was a nation of small towns, farms, and villages. Although the great cities were in place, they did not hold then the overwhelming percentage of the population that they hold now. For most of the people, the stars were the principal lights of night, familiar and clear and bright.

But today America is a massive web of electrified light, and the cities pour out so much radiant pollution that an urban dweller can make out at best only a few of the stars that elsewhere jump and glitter in the sky. It is against this unremitting assault of street lamps, house lights, and neon that Halley's must compete. Without help from below, there is every chance that the comet will fail, and that viewers caught in the midst of light pollution will simply be unable to see the comet, or be able to see only a pale image of it.

Now, help from below is materializing.

Fred Schaaf, a columnist for *Astronomy,* has founded an organization called Dark Skies for Comet Halley. The aim of the organization is to persuade municipalities to douse the lights for the few consecutive nights when Halley's will be best positioned for viewing. Dark Skies anticipates a lot of resistance in its campaign to give Halley's a properly darkened stage, but perhaps they needn't be as pessimistic as they are. At least one major city has already demonstrated that—if the arguments are compelling—it is prepared to suffer a little less light.

In a victory for astronomers at the nearby Palomar Observatory, the city of San Diego has agreed to install a new kind of streetlight that astronomers say will make their viewing easier. Scientists from the California Institute of Technology, which operates the prestigious facility, say the municipal decision is a major victory against light pollution in the United States.

Before the city made its decision, nearly a third of its 27,000 streetlights had been converted to high-pressure sodium lamps, which the astronomers claimed cut seriously into the light-collecting abilities of the telescopes at the observatory, sixty miles north of San Diego. The city had not made its decision lightly. The new high-pressure lamps were saving it $1,000 a day on its light bill. But in the end the municipal councilors felt it was better to preserve a resource like Mount Palomar than to save the money.

Had the city continued to install the high-pressure sodium lamps, the astronomers said, and had it continued to grow at its present rate, then even the massive 200-inch telescope atop Mount Palomar would have been blinded by the encroaching light within fifteen years. But the *low*-pressure lights operate in a single band of the spectrum, and the scientists are able to work around this. Robert Brucato, Palomar's assistant director, has likened the low-pressure light to a picket fence with a single picket . . . easy to see through. The high-pressure light, he said, was more like a brick wall.

Rehearsal on the Mountain

In the spring of 1984 a group of astronomers assembled on Table Mountain, a 7,500-foot summit in the San Gabriel Mountains northeast of Los Angeles. The astronomers were there to refine their techniques for comet watching, so that when Halley's comes belting in on the last leg of its orbit around the Sun they will be ready for it. It was a cool April evening when the astronomers gathered around the stout twenty-four-inch telescope maintained at the mountaintop observatory by the Jet Propulsion Laboratory in Pasadena.

What they were all straining to see that night was a blurred little patch of light originating from an object 74 million miles out in space: Comet Crommelin. Crommelin is a small, dim comet, not even close to the magnificence of Halley's. The gathering atop the mountain that night was just part of an astonishing global arsenal of astronomical talent being marshaled to study Halley's. Virtually every telescope in the world will be trained on the famous visitor, as well as a battery of scientific instruments unknown when the comet orbited the sun in 1910 on its last visit.

To create as detailed a record as possible of the comet's passage, a worldwide network of observers has been established under the aegis of the International Halley Watch. The Halley Watch will be directed from the Jet Propulsion Laboratory's premises in Pasadena and from the University of Erlangen in West Germany. There will be more than 800 astronomers representing forty-seven countries taking part in the program to track the comet and monitor its every change. In addition, hundreds of amateurs have volunteered their services to assist the professionals. From every point on the globe, amateurs will be training their instruments on the comet and relaying their observations to the scientists who are readying to receive the massive input of data.

That night on Table Mountain, three astronomers were assembled to practice on tiny Comet Crommelin. James Young, the mountain's resident astronomer, peered critically into the eyepiece of his instrument. There was a lot of turbulence in the atmosphere, and Young was skeptical about the quality of the viewing available. But he found the elusive speck of light at last, and locked on. As he squinted through the eyepiece, it was easy for Young to see why Crommelin was difficult to locate. Like a sparrow in a windstorm, the comet jumped and danced in the turbulence.

Although the early views of Halley's will be like those of the skittish Crommelin, the later views will be more satisfactory. A good deal larger and several hundred times brighter than Crommelin, Halley's will also sport a magnificent dust tail millions of miles long. But the scientists on Table Mountain had to live with Crommelin, and they set to work.

As Dr. Stephen Edberg began to operate the controls that focused the powerful instrument on the distant object, Dr. Jay Bergstralh commenced to take a series of measurements with a device known as a photometer. The photometer is fixed to the telescope's base, and is designed to detect shifts in the intensity of light radiating from distant objects. Says Dr. Bergstralh: "There's very little visual observation done by professional astronomers, and getting to be less photography too. The telescope mirror is just a big light bucket. It collects and we interpret the light with various detectors like the photometer."

By directing the incoming light through different filters, the astronomers hoped they would be able to identify separate molecules blasted from the comet by the torrential solar winds as it neared perihelion. Other measurements may even allow the Earth-bound scientists to discover what makes up the cometary nucleus.

As the astronomers stood in the chill April air, the information collected by the photometer recorded onto computer tape and, at the same time, chattered out a printout on a teletype machine attached to the telescope. The astronomers will tabulate all the new data, comparing them with data from well-known stars collected at the same time. Knowing the way these familiar stars behave in different atmospheric conditions, the scientists studying the comet will be able to make allowances for atmospheric effects on the cometary data.

Just as the astronomers at Table Mountain were feeding their new information into the Halley Watch computers at the Jet Propulsion Lab, data from astronomers posted around the world were also pouring into Pasadena. Every one of the astronomers had been watching Crommelin from his own post. Computer to computer, telephone to telephone, telex to telex, the information rattled into the California headquarters. The purpose of the exercise was to ensure that everyone taking part in the Halley Watch was using the same standards of reporting and observation. The directors in Pasadena also wanted to test their communications channels with the rest of the world. As Ray Newbury, Jr., the Halley Watch leader in Pasadena, put it, the exercise "allows us to flex our muscles and see if we have muscles, or whether it's all flab."

The Pioneers

According to Dr. Edberg, the biggest gaps in observational coverage of the Halley's flyby will be the oceans, particularly the tremendous expanse of the South Pacific.

"Fortunately, we have some observers who are going to be in some of the island groups at the right time, but that's one awfully big ocean down there and there's just no way we can fill in the hole as well as we'd like."

Despite its sophistication and considerable resources, the International Halley Watch still depends upon people to make it work. The exotic telescopes will not do it themselves. And, in fact, the breathtaking astronomical observatories that we take for granted today are fairly recent adjuncts to the science of astronomy. Plans have already been laid for the staggeringly powerful 400-inch Keck telescope to be built on Mauna Kea in Hawaii. There, the Keck installation will join a clutch of other observatories already in residence. But it was a very different Hawaii that greeted the first American astronomer to use the islands for his observations.

What may have been the first high-altitude observatory in the world was built on Mauna Loa in 1840 by Lt. Charles Wilkes of the United States Navy. (Mauna Loa has since become active but was not then.) The Wilkes expedition sailed from Norfolk, Virginia, in 1838, entrusted with the mission of making observations to determine the exact longitude and latitude of certain Pacific islands. When he landed at Hilo on the island of Hawaii in 1840, Wilkes first tried to establish his observatory near the shore, where it could be easily serviced by ship. But the saltwater spray thrown into the air by the heavy surf was a serious impediment to observations, and Wilkes decided to climb the nearby mountain.

It was December when Wilkes and his party set out to ascend Mauna Loa. At that time of year there was snow on the higher reaches of the mountain, and when they reached the first of the snow the Hawaiian bearers promptly deserted. But the Wilkes party forged ahead and, when they had gained the summit, erected a small observatory.

As well as making the observations of latitude and longitude that he had come for, Wilkes also made some observations of Jupiter's satellites. Planetary astronomy was the popular astronomy of the day when Wilkes made his observations, and the data on Jupiter's moons that he brought back to the United States excited a good deal of interest among the public. In fact, W. R. Beardsley of the University of Pittsburgh believes that it was the interest generated by the Wilkes expedition that led to the founding of both the Harvard College Observatory and the U.S. Naval Observatory.

It was also to Hawaii that the American Astronomical and Astrophysical Society decided to send an expedition in 1910, when public excitement at the apparition of Halley's comet had turned the event into a carnival of feverish speculation and interest. But the astronomers who man the International Halley Watch today would be appalled at the complete failure of support for the 1910 expedition. The astronomers could get no money. They could get no personnel. At last the National Academy of Sciences coughed up the funds. George Ellery Hale, the great astronomer for whom the Hale Observatories are now named and who was then the director of the Mount Wilson Observatory, was asked to find someone to lead the expedition. Hale asked Ferdinand Ellerman, who accepted.

Ellerman gathered his troops and, loading all his people and gear aboard a ship, sailed off to Hawaii.

The expedition established itself on Oahu and set up its equipment on the beach close to the south slope of Diamond Head. The main objective of the expedition was to take a series of wide-angle photographs of the comet's tail. Seventy photographs were taken between April 14 and June 10, 1910. Although some of these photographs were ultimately published, there was little interest among astronomers, and most of the scientists of the day considered that the expedition was a failure.

Of course, it could hardly have succeeded. Compared to the detailed plans that astronomers have laid for the present appearance of Halley's, poor Ellerman was simply cast adrift. The astronomers of the day—including Ellerman—had no particularly clear idea of what they ought to be looking for in a comet in the first place. Since then, the development of theories about cometary composition, and the development of equipment to test those theories, have given scientists much more direction than was possible when Ellerman and his expedition set out for Hawaii.

In Comfort

Like Ellerman and, before him, Wilkes, there are modern observers who plan to set sail, too. But they will do so in a lot more comfort. By late 1984, at least one cruise line had made plans to carry comet watchers to sea. The idea is not a new one. Several large ocean liners carried enthusiasts out of New York for cruises devoted to Halley's in 1910. And with the light pollution of our present day to contend with, the sea will probably be the best place to get an undiluted view of the comet.

The Sun Line has planned two cruises, both for 1986. The *Stella Solaris* and the *Stella Oceanis* will be sailed out to sea and positioned at predetermined coordinates that should provide passengers with the best vantage points for viewing the comet. The cruise line operators plan to augment their regular staff with a team of astronomers and others with specialized knowledge, who will deliver lectures on the history and science of comets and also give the passengers practical advice on how to photograph the bright visitor.

The *Stella Solaris*'s fourteen-day cruise begins in Manaus, Brazil, on January 4, 1986. Passengers will fly from Miami to Manaus. The ship will bear her cargo of comet gazers into the South Atlantic to view Halley's in southern skies as the comet makes its closest approach to Earth. For the present apparition of Halley's, the more southerly latitudes are the most favorable for viewing the comet. It is impossible to predict accurately, but in ideal conditions a viewer in Southern latitudes should be able to see a tail anywhere from thirty to sixty degrees long. The actual length of the tail will be many millions of miles.

The *Stella Oceanis* will begin her cruise on March 15, 1986. Lasting thirteen days, the cruise will carry passengers to a number of Caribbean ports. At that time of year the comet should become visible in the morning sky just before sunrise, becoming brighter from day to day.

Cashing In

In 1910 there were the comet pills and the comet helmets, both designed to protect the helpless Earthling from the perils of the comet's passage. There were plenty of entrepreneurs around in 1910, and there are plenty now. Looking back on the marketing adventures of his predecessors of seventy-five years ago, Burton

Rubin believes that "the world hasn't changed that much," and he's banking on it. Rubin is a New York entrepreneur, and the founder of a company called Halleyoptics.

"First there were the Democratic and Republican conventions. Then there was the Olympics. Then there was the election. And now everyone will be concerned with the coming of Halley's comet," Rubin proclaims in his New York office.

Mr. Rubin is a firm believer that the present apparition of Halley's will generate the same mix of curiosity, trepidation, and awe as it did when it swept through the skies in 1910. But anybody can be a believer. Being an investor takes a different kind of faith. Mr. Rubin has that kind of faith, and so he has Halleyoptics.

"I have always been interested in the madness of crowds," Mr. Rubin affirms.

Halleyoptics manufactures the Halleyscope, a lightweight telescope that also performs double duty as a telephoto lens for a camera. Mr. Rubin hopes that enough comet watchers will turn to the Halleyscope for their viewing to make him a fortune. It will not be his first.

Borrowing $5,000 in 1968, Mr. Rubin and a partner, Robert Stiller, founded the E-Z Wider Company. E-Z Wider specialized in the manufacture of double-width rolling papers for smokers. "Double-width" refers to the thickness of the paper. You may wonder why smokers suddenly required papers of unusual toughness when the ordinary variety had served so well for so long. The answer is that most of these smokers were not smokers of tobacco, but smokers of marijuana, of whom there were an estimated 40 million around the world. When he sold out in 1980, Rubin pocketed a profit of $6 million. With his security assured, the entrepreneur began to take more time off, spending it with his son. Together, they would head off for the country and look at the stars. Something began to click and whir in Rubin's fertile brain.

Digging through the records of newspapers like the *New York Times* and the *Christian Science Monitor,* the entrepreneur began to assemble a file of stories about the antics of people during the last apparition of Halley's. Many of these were alarming—the tales of lunacy and misadventure tied by popular superstition to the appearance of the comet. But other stories were cheerful. Many Americans had, Rubin learned, greeted Halley's with exuberance, throwing parties, arranging picnics and balloon rides, and piling into great hotels like the Plaza in New York, where bartenders provided a concoction known as the Comet Cocktail.

An idea took shape as Rubin read on, and that idea was the Halleyscope. What finally decided the entrepreneur was the boast of one New York vendor of 1910 who claimed that there were more telescopes sold in the three months before the comet showed up than had been sold in the whole period since the Civil War. If Rubin's plans go aright, the same thing will happen this time.

"You don't have to be an expert to use the Halleyscope," Rubin maintains. "It is designed for use by anyone who can operate a thirty-five millimeter camera."

The Halleyscope weighs less than a pair of binoculars, and rests on a tabletop tripod. In 1984 the scope was selling for about $250. Rubin believes that real comet fever will strike the world about September of 1985, and he says that a Japanese factory has 100,000 of the scopes all ready to ship in expectation of a massive surge of demand.

"In 1983, 632,000 amateur telescopes were shipped to the United States from Japan," Rubin claims, citing this as support for his belief that "there is no better time to start an optical company."

Together into Space

On a more lofty plane than the spirited cupidity of the entrepreneur is the mutual generosity developing among the international scientific community as the comet swings onto final approach. When the $75-million Giotto spacecraft being launched by the European Space Agency heads into the last stages of the delicate maneuvering that will direct it to within a few hundred miles of Halley's, in March of 1986, the last key course corrections will be based on information supplied to the Europeans by the Soviets. That exchange will be just one of the results of an exceptionally bountiful exchange of data among the Europeans, Soviets, Japanese, and Americans. The aim of the space scientists is to boost the chances for success of *all* the missions to the comet, rather than just the missions of their own countries or groups.

Dr. Carl Christiansen, a member of the Jet Propulsion Laboratory, said at a meeting in Moscow that the collaboration was the "most ambitious piece of Soviet–American cooperation since the Apollo–Soyuz linkup of 1975." Dave Dale, a Giotto program manager, agreed, expressing relief that a formal agreement among all the nations participating had finally been signed. "It took eighteen months to get it through the bureaucracy," Dale said.

The Russian mission to the comet began in December of 1984, with the launch of two Vega spacecraft destined first for the planet Venus. There, the Vega craft will drop balloon probes into the cloudy Venusian atmosphere and move on. Their next destination is Halley's.

To help the Russians plot their course for intercepting the comet, the United States will assist in tracking the cometbound Vegas. The American deep-space network is capable of much more accurate fixes than is the Soviets' ground-based radar. And to make certain that the space network is perfectly attuned to its new task, American space engineers have spent $2 million in up-to-the-minute modifications.

Finally, the Vegas will close to within 6,000 miles of the comet in March of 1986. Back will pour a flood of what scientists call hard copy, reams of cryptic data that must be fed into computers and decoded in order to be made comprehensible to the vast majority of Earth-bound watchers. Much of this information will be instantly relayed to European space controllers, who will feed the last-minute course corrections to Giotto.

"It should increase our accuracy from seven hundred kilometers to one hundred twenty," said Dale.

Early in 1985, only five months before its scheduled July launch, Giotto was subjected to a thorough battery of tests at the European Space Agency's test station in France. Engineers placed the British-built Giotto in a vacuum chamber, and there it underwent a series of violent spins and shakes and was subjected to the extremes of cold and heat that it will encounter, first in space and later as it nears the great comet itself. But all of this vigorous shakedown is only to get the probe up *to* the comet. There is no chance that it will survive its close look and return to Earth.

For, after relaying her own hard copy back to the Europeans' receiving stations, measuring the chemistry and magnetism of the comet's tail, and after it has taken and sent its breathtaking pictures back to its masters millions of miles below, then Giotto takes its last, heroic plunge into mystery. Turning into the stream of dust and plasma pouring from the rocketing comet, Giotto will strike straight into the ever-denser regions of the tail, making for the nucleus itself. Long before it reaches that beguiling lump of rock and ice, Giotto is expected to be torn apart by bulletlike particles of dust. So high is the velocity of these cometary motes that the effect of one striking the comet would be roughly equivalent to an automobile rushing at 100 miles an hour into a sheet of glass.

EPILOGUE

Finding the Comet

There is good news for Halley's watchers. The initial predictions that the comet will be difficult to see have at least been softened. Two researchers working out of the Prospect Hill Observatory in Harvard, Massachusetts, believe that our visitor may be as bright as Polaris. Your newspaper or local observatory will be able to give you exact directions, but in general here's where to look.

In December of 1985, the comet will be visible to anyone with binoculars between the horizon and the zenith (directly above you) in the southwest about an hour and a half after sunset. By early January 1986, you should be able to see it with the naked eye. During January the comet will have brightened and developed a tail as it approaches the Sun. Its position in the sky will move lower and more toward the west.

By the end of January, the comet will have disappeared into the solar glare, and will be lost until it appears again a month later, this time in the *morning* sky.

A VIEWER'S GUIDE TO HALLEY'S COMET

Look for the comet an hour and a half before sunrise low on the horizon and just south of due east. As March progresses, the comet will ascend the sky and move to the south, brightening all through the month. Late March and early April are the best times for looking at the comet. By then it will have a handsome tail and be giving us its best show. By late April the comet will have shrunk and moved into the southeast, once again in the *evening,* an hour and a half after sunset.

By May it will be well on its way to where it came from, and we can all put away the comet pills for another seventy-six years.

Comet Talk

Aphelion: The point of the comet's orbit that is farthest from the Sun.

Apparition: Astronomers' word for a celestial appearance. The length of time that the comet is visible from Earth.

Coma: The great, luminescent cloud of gas and dust surrounding the nucleus of a comet.

Dust tail: Solid particles of dust blasted from the surface of the comet's nucleus by the pressure of the Sun's radiation. The effect of streaming fire is partly caused by the reflected sunlight striking these particles.

Head: The nucleus and coma of a comet, taken together. In the case of a comet such as Halley's, the head may be hundreds of thousands of miles in diameter.

Ion tail: The name given to the tail that is formed when the solar wind surges into the comet, dragging off it the electrified gas that forms the ion tail. The light from the ion tail is caused by fluorescing ions.

COMET TALK

Nucleus: The heart of the comet, the matter at the center of the mighty show. Current theory suggests that the nucleus is made of water ice and rocky compounds. Halley's nucleus is probably only a couple of miles in diameter.

Perihelion: The point of the comet's orbit that is closest to the Sun.

Solar wind: Ionized gases pouring off the Sun at velocities of more than a million miles an hour.

Tail: Collectively, the dust tail and the ion tail. The tail comprises all the ejecta strewn out behind the comet, and may be as long as 50 million miles.

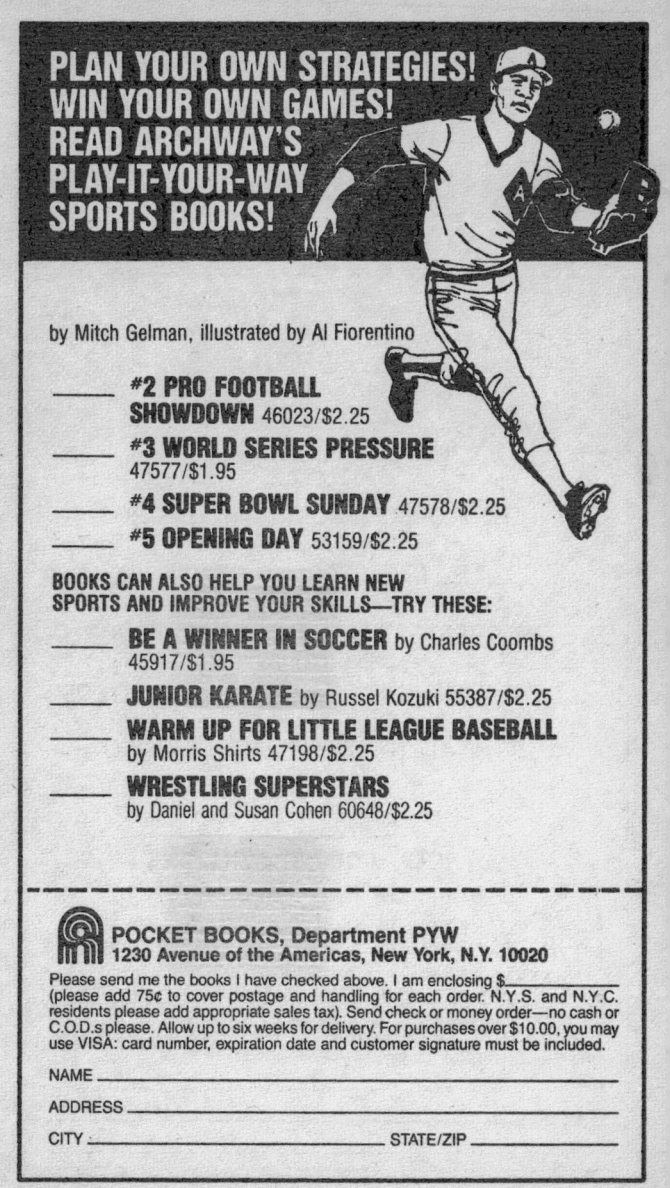